MARY ANNE'S GARDEN

to become a hero;
One often dreams
of it, or perhaps
'tis only I who
dreamed of it as
a child and yet,
alas, never have
I had the
opportunity
until this
(well, a sort of hero
thwarted bud
reach full
slipping away
to watch
spring   into a
midday maturity
As the vine moves on
As the vine always

morning
to allow this
to unfold and
Slowerly simply
the vine and
the slow motion
whole and round
within the hour

moves on ...

august eighth

MARY ANNE'S GARDEN

DRAWINGS AND WRITINGS
BY MARY ANNE MCLEAN

HARRY N. ABRAMS, INC.
PUBLISHERS, NEW YORK

TO ANNE DE BEAUMONT MCLEAN
For the growing good of the world is partly dependent on
unhistoric acts; and that things are not so ill with
you and me as they might have been, is half owing to the
number who lived faithfully a hidden life, and rest in
unvisited tombs.—George Eliot, *Middlemarch*

AND RUTH MCVAUGH ALLEN
She was that different drummer

I wish to express gratitude to Peter Bradford for the
land and the leisure which made these drawings possible.

Project Director: Robert Morton
Editor: Ruth A. Peltason
Designers: Peter Bradford and Lorraine Christiani

Library of Congress Cataloging-in-Publication Data
McLean, Mary Anne.
    Mary Anne's garden drawings & writings.
    1. Gardens—Pictorial works. 2. Botanical illustration.
3. Vegetables—Pictorial works. 4. Pastel drawing, American.
I. Title. II. Title: Mary Anne's garden drawings and writings.
SB450.98.M376   1987   635'.022   87-1016
ISBN 0-8109-1418-2

Times Mirror Books   Printed and bound in Japan

CONTENTS

There are some things which in themselves demand a physical reaction. Try not to hold a conch shell up to your ear for the sound of the sea. Try not to stroke a beautiful animal. Once in Mexico I cut my long hair very short; the land was so exhilarating I had to do something. I couldn't stroke it or hold it up to my ear and if I could sing maybe I wouldn't have had to cut my hair.

I felt very much the same about the young lettuces. The pale leaves came up with such grace and eloquence, were so dance-like in their debut that I couldn't get them close enough to me. By tracing the motion of their lives perhaps I could gather for myself a bit of their dance. It has been said that what man really wants is to be at one with nature. I think this is true. I wanted to be at one with lettuce and that is why these drawings began.

But the wondering came first. One might expect it to have begun in the rosy hours of the morning, with a willow basket over my arm, pondering the mysteries of my garden. Rather it began in the cold light of the A&P by a rotating seed display that had been placed beside the lettuce bin. There, side by each, the lettuce and the lettuce seed. I stood dumbstruck. This lettuce and this seed had something to do with each other. I knew, of course, that the seed made the lettuce. But I had never once wondered how the lettuce made the seed. I suddenly needed to know. I decided to buy a packet of seeds, plant a few, and watch very carefully.

In the days and weeks that followed I watched and drew the lettuce at different stages. I enjoyed the pleasant rhythm of the watching and the drawing. I became so excited when I discovered how the lettuce makes the seed that I wanted to explore other vegetables and flowers. I began to draw the growing things I was curious about, not knowing I was beginning a ten-year chronicle.

When the plants in my garden were very small I would lie, chin to earth, to draw them. As they grew taller I'd sit on the big soup pot I carried outside; turned upside down it was a substantial seat and gave a pretty good height for observing most things. As I drew I became aware of animals I had never noticed before. And as they'd appear among the plants I'd draw them as well.

In making my pastel drawings I purposely didn't compose the pages, not wanting to be intimidated by the suitable-for-framing approach. I wrote down my thoughts at the time, usually observations on the subject I was drawing, but often just ideas which would spring to mind. Since these studies were not intended for anyone else to read, it didn't matter how my thoughts drifted in and around my subjects. Each drawing answered some questions for me, but posed others. Yes, the cabbage worm eats from his right to his left, and his jaws open and close sideways like little hedge clippers, but just how much does he eat in a day, how much does he rest, and where?

To say that *Mary Anne's Garden* is a series of time-lapse drawings of garden things is quite true, but oh my, that sounds like a bit of a yawn. To me the drawings are the surprises and excitements of things growing, the seeing of them as not separated from one's own growth, the stepping a little closer to one's self for stepping closer to them.

THE VEGETABLE GARDEN

What brings me out to the vegetable garden every other hour? It isn't the weeding or watering or even the drawing. Is it the parent in me checking the child in its crib?

I just like the feeling of the place. I drift out and stand here. What do I think about? Everything at all! I notice some things I hadn't noticed before or I wonder about something: The veins on the underside of an eggplant leaf are as soft, as curved and carved as those on the back of my hand....The sun is shining a glove on my hand....Vegetables are as variable in shape and color and texture as the ornaments on a Christmas tree, and are as crisp and savory as a holiday.

If a snow pea tendril moves too slowly for my eye to register, do I move too fast for the plant to register me at all? Like the swift comic-book hero shown as a rush of lines where he has been, am I just the rush of lines of where I've been?

I like to stand in the shelter created by the tall pole limas. The simple rectangle of rows and aisles intrigues me. How did man think of parallel lines in the first place? Where do they appear in nature, in a stand of trees, or fingers, I suppose, and toes? This classic vegetable garden format has remained unchanged throughout countries and centuries. I like this continuity.

I go inside to homely tasks.
I am back in two hours.

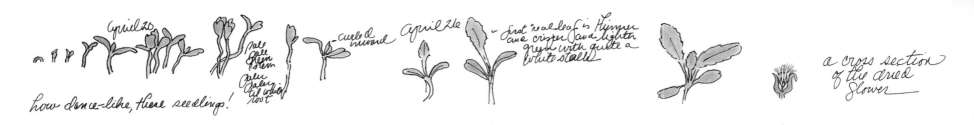

April 20

pale pale green stem

new palery tip white root

curled inward

April 26

first "real leaf" is thinner and crisper and lighter green with quite a white stalk

a cross section of the dried flower

how dance-like, these seedlings!

the first few leaves seem very loosely gathered together

After a short while, the new leaves know how to turn in and a firm "head" is formed. the new leaves always forming from the inside.

If the head is not harvested, the lettuce will go to seed. First, the stem, which was not even obvious existed, extends itself drastically. At the point of this drawing, the plant is 2 feet tall (I show only the top foot) and the leaves have lost their chartreuse translucent quality to become minty green and opaque.

Now three and a half feet tall, the tip has branched out extensively and holds many yellow blossoms which will eventually dry to a white fuzz. Each seed is attached by the very finest filament to a section of fuzz.

11 Lettuce

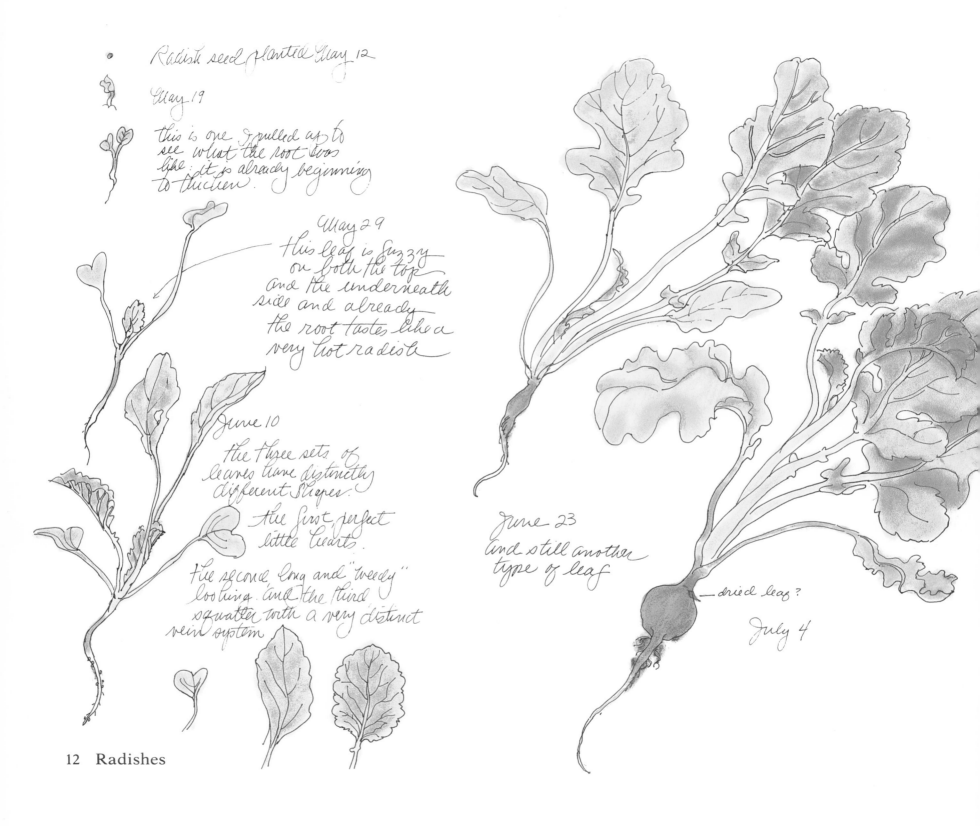

Radish seed planted May 12

May 19

this is one I pulled up to
see what the root was
like; it is already beginning
to thicken.

May 29
This leaf is fuzzy
on both the top
and the underneath
side and already
the root tastes like a
very hot radish

June 10

the three sets of
leaves have distinctly
different shapes.

the first perfect
little hearts.

the second long and "weedy"
looking. And the third
squatter with a very distinct
vein system

June 23
and still another
type of leaf

dried leaf?

July 4

13

July 11

July 28

This is the full plant, actually about 40" tall. The pods have grown very lumpy with a bit of magenta at the base of each. Now I know this has nothing to do with radishes but it is 6:00 in the evening and two birds are calling back and forth and they are saying "tweet tweet." Well now really that is just a caricature of a bird song. I always thought that "bow wow" was such baby talk and I looked upon it with a certain disdain, then I realized that Shakespeare said it and probably made it up. So maybe he made up the "tweet" as well.

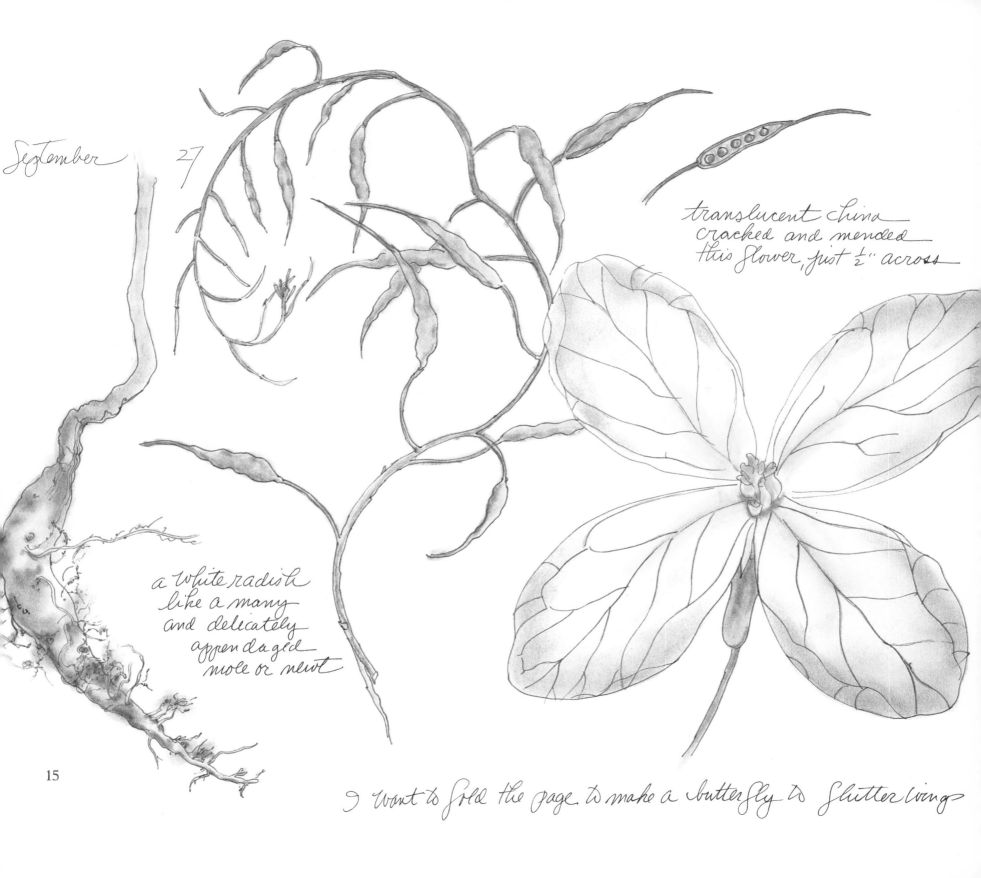

September 27

translucent china
cracked and mended
this flower, just ½" across

a white radish
like a many
and delicately
appendaged
mole or newt

I want to fold the page to make a butterfly to flutter wings

Snow Peas

May 20

May 29

dewdrop

(In just 3 days this "butterfly" folded leaf complex unfolds & extends itself startlingly — the tendril here.)

June 5

June 8 is now inside that bottom "butterfly wing" is the whole thing all over again and again.

there on the main stem are quite like water lily leaves and perhaps a bit more chalky or minty in color also, silvery spotted.

very large opened "butterfly" from another larger plant

and when I pull this outter leaf aside these two units are inside and the leaves are very translucent

Should I keep unfolding and unfolding these layers of leaves I would find them like little wooden Russian peasant dolls each one inside the other, but smaller and smaller and smaller

June 10

Just realize that the water lily leaf alternates on the stem first directed one way then the other and that the stem is four sided

eaten?

minetwists,

June 14

The tendrils had lost their grip on the fence and the vine had fallen down. I helped it back up again and returned it to the same spot on the fence. Someone has nibbled a bit of leaf. This is a beautiful cool and sunny day. I hear crows calling back and forth.

16 Snow peas

I always love
these little
holding hands
tendrils

July 14

July 15   true true
a bee did pollenate
this blossom while
I was here

and the very
next day, the
pod has grown
this much.
the blossom
has dried up and
barely hangs on.
Little white silks which
must be the stamens
hug the sides of the pod

July 17   the pollenated
blossom has withered

perhaps this blossom
had been pollenated
before because could it
really, in two days have
a pod developing? I've
probed most gently so
as not to disturb the
natural timing, but there
is the tiniest pod inside

and also July 17, the blossom
has dropped from the above pod

a bit further down than this, the stalk
has become cracked and weak and
since I have come to depend upon this
very vine for my sequence I have bound
and supported the wound with a bit of tape

July 17

July 19

Does the tendril say
"watch me, watch me"
to the mother plant
as does my child when
going down the
sliding board?

Development of the snap pea and

the "smile"

18   Snow peas and snap peas

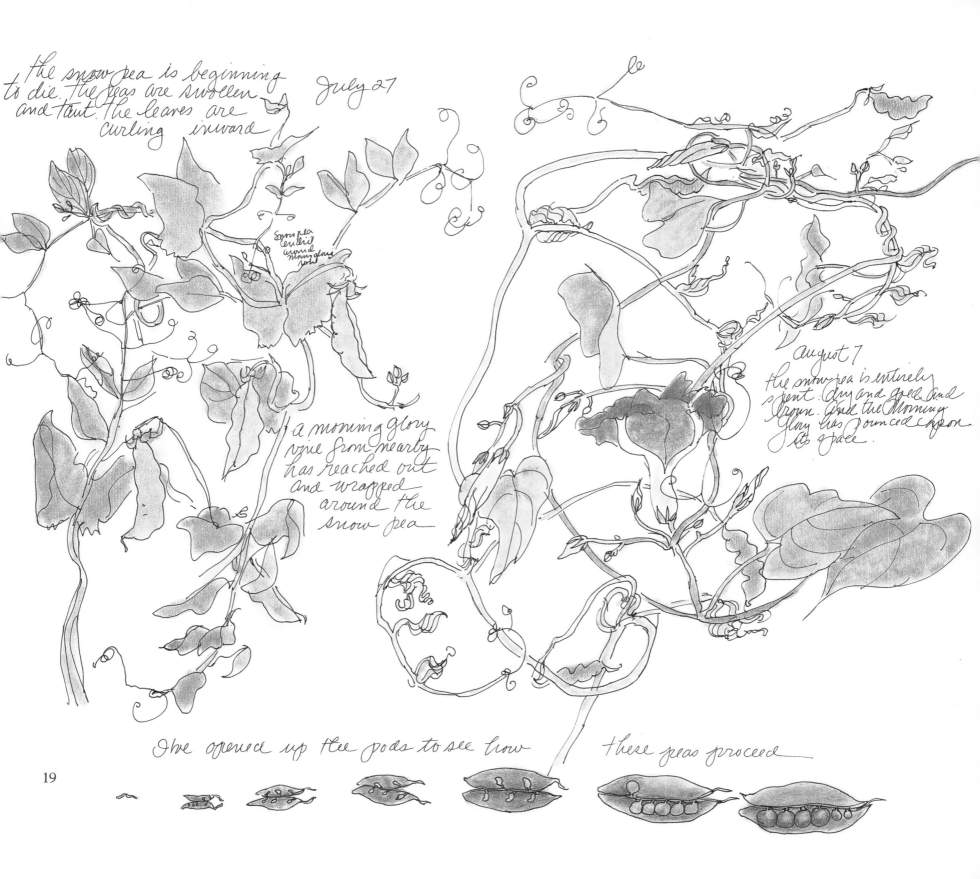

the snow pea is beginning
to die. The peas are swollen
and taut. the leaves are
curling inward

July 27

Snow pea
tendril
around
morning glory
leaf

a morning glory
vine from nearby
has reached out
and wrapped
around the
snow pea

August 7
the snow pea is entirely
spent. dry and yellow and
brown. And the Morning
Glory has pounced upon
its space.

19

I've opened up the pods to see how   these peas proceed

I planted this row of string beans May 25. Today is June 5, just ten days later and already the row makes a definite stripe in the garden. I think of the garden sometimes as a flag, all those wonderful tidy green stripes and some day I fancy I shall not be able to keep myself from adding a flagpole a reclining one in the spirit of "Christo". Isn't it like some great little half-time band formation? And also like a musical manuscript? All those lovely lines and dots and curves keeping some sort of quiet cadence? Do you suppose that Mozart thought his manuscripts looked like gardens? I saw once a piece written in his own extraordinary hand with some notes in green ink, some in red. But oh, and oh, this garden is as close to writing music as I shall come. Now these little seedlings are all on this day in varying stages of development and I can see changes occurring from the time I started the drawing to the finish time so I shall attach myself to the seedling which is fourth from the left and which at 1:30 still had the tip of its leaf in the soil. Now, at 2:30 the tip is quite free, the "trunks" are vaguely fuzzy: the underside of the leaves are a silvery mint with light veins.

2:30 P.M.   3:15 P.M.   7:00 P.M.   8:30 P.M.   8:00 A.M.   12:00 P.M.   3:00 P.M.   7:00 P.M.

20   String beans

On an orchestra, all the various texture and tones, the orderly seating arrangement all the strings together and the woodwind .... The sun being the obvious conductor and more like the orchestra at Radio City Music Hall than any other, coming up out of the ground the way it does ____ But enough enough!

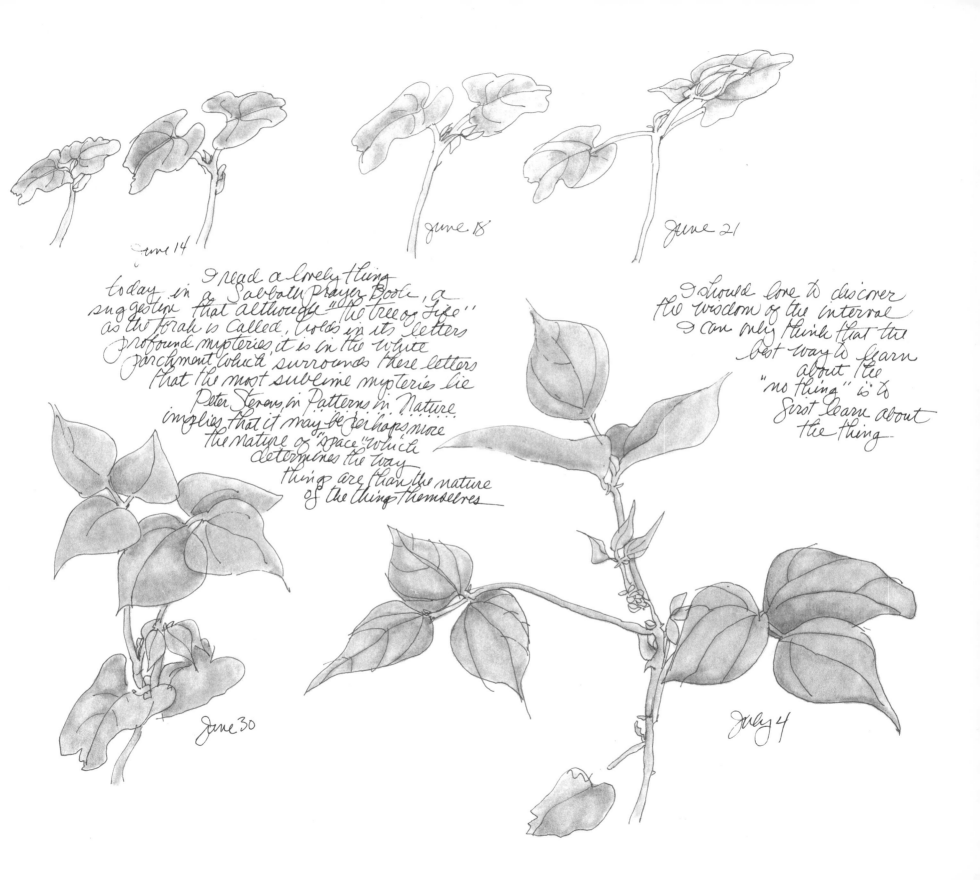

June 14

June 18

June 21

I read a lovely thing today in a Sabbath Prayer Book, a suggestion that although "the Tree of Life" as the Torah is Called, holds in its letters profound mysteries, it is in the white parchment which surrounds these letters that the most sublime mysteries lie
Peter Stevens, in Patterns in Nature . . . . . . implies that it may be perhaps more the nature of "space" which determines the way things are than the nature of the things themselves

I should love to discover the wisdom of the interval
I can only think that the best way to learn about the "no thing" is to first learn about the thing

June 30

July 4

July 17

bud   flower   bean.

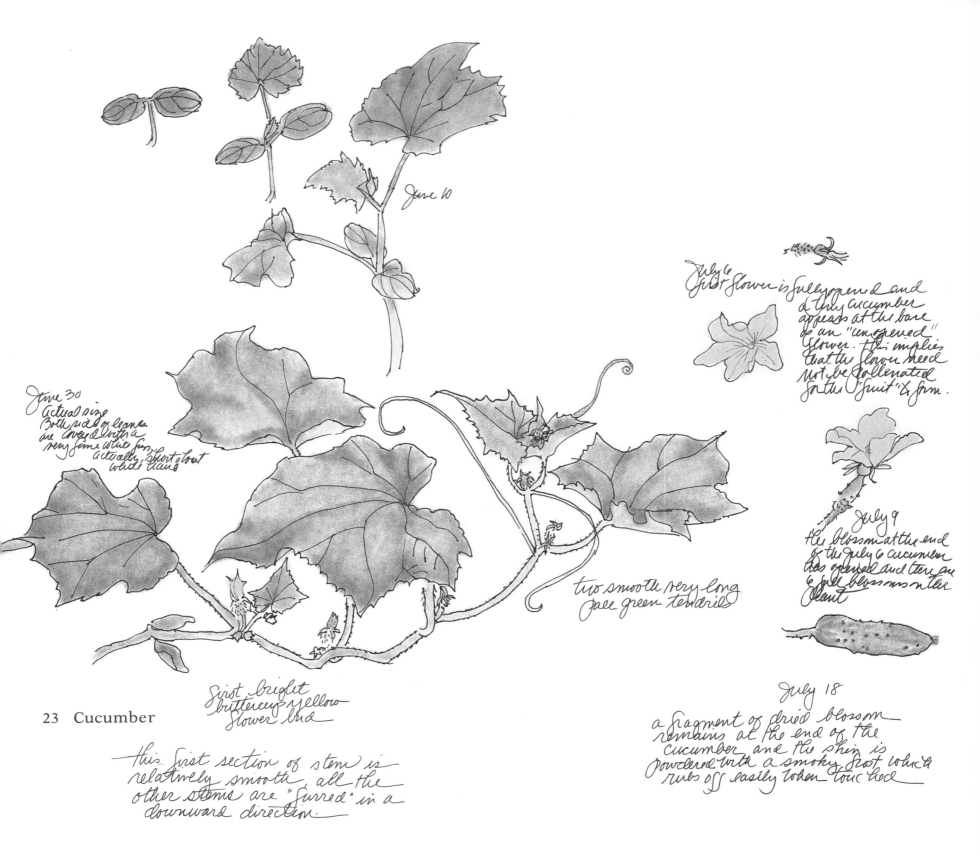

June 10

July 6
First flower is fully opened and a tiny cucumber appears at the base of an "unopened" flower. This implies that the flower need not be pollenated for the "fruit" to form.

June 30
actual size
Both sides of leaves are covered with a very fine white fur, actually short stout white hairs

two smooth very long pale green tendrils

July 9
the blossom at the end of the July 6 cucumber has opened and there are 6 full blossoms on the plant

23 Cucumber

First bright buttercup yellow flower bud

July 18
a fragment of dried blossom remains at the end of the cucumber and the skin is powdered with a smoky frost which rubs off easily when touched

this first section of stem is relatively smooth, all the other stems are "furred" in a downward direction.

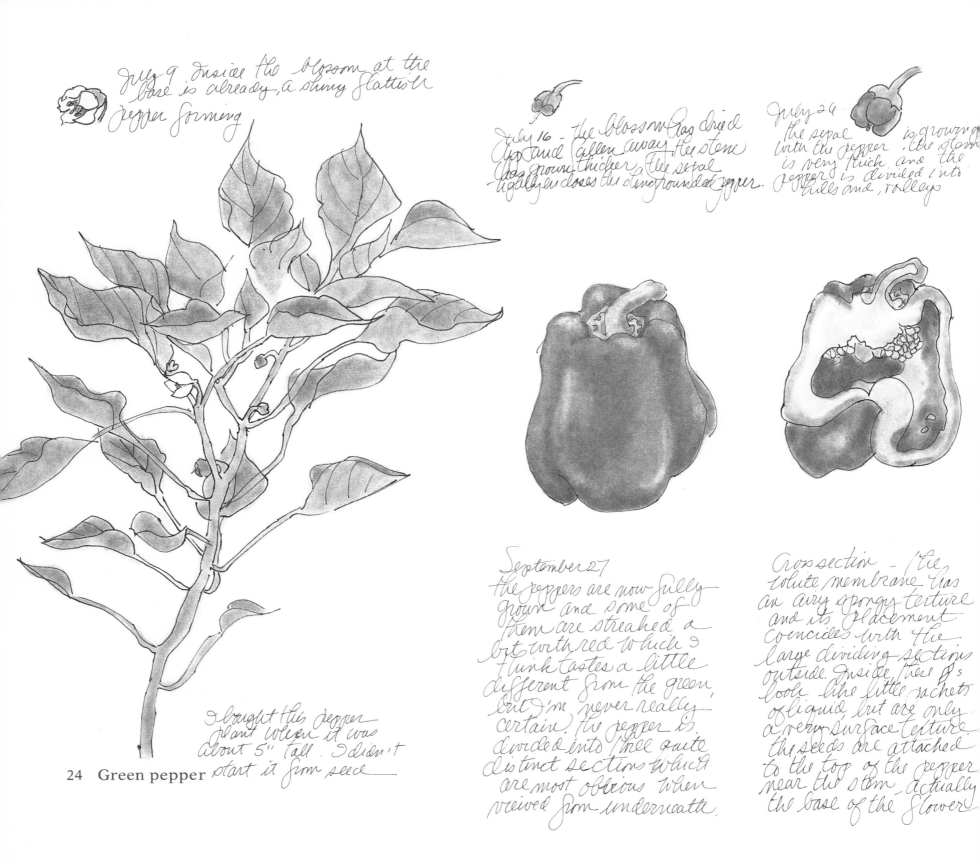

July 9 Inside the blossom at the base is already, a shiny flattish pepper forming

July 16 - the blossom flag dried up and fallen away. the stem has grown thicker, the sepal tightly encloses the tiny rounded pepper.

July 24 the sepal is growing with the pepper. the stem is very thick and the pepper is divided into hills and valleys

September 27 the peppers are now fully grown and some of them are streaked a bit with red which I think tastes a little different from the green, but I'm never really certain. the pepper is divided into three quite distinct sections which are most obvious when viewed from underneath.

Cross section - the white membrane has an airy spongy texture and its placement coincides with the large dividing sections outside. Inside, these B's look like little sachets of liquid, but are only a very surface texture. the seeds are attached to the top of the pepper near the stem - actually the base of the flower

I bought this pepper plant when it was about 5" tall. I didn't start it from seed

24   Green pepper

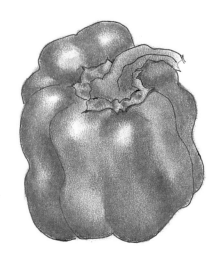

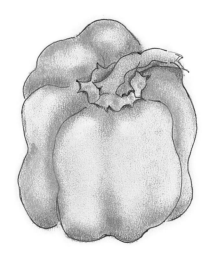

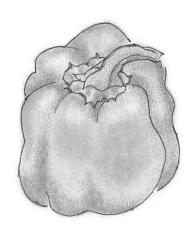

October 1 - The pepper is
very bright and full of
highlights with a sharp
crisp smell. It sounds
hard and hollow when
you knock on it.

October 8 - Everything has
mellowed. the color has
changed to a soft yellow-
ish translucent green. It
feels soft and sounds soft
and the smell is much
subdued.

October 15 - Quite a bit
of orange color has
appeared. the pepper
has gotten smaller and
softer and it is begin-
ing to wrinkle.

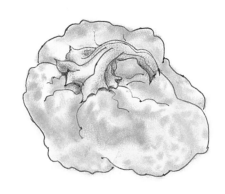

October 22 - the skin, very
soft, wrinkled and bright
orange - just touches of green
remaining. the stem is
dry and is separating from
the body of the pepper.

October 27 - Decidedly
collapsed now and with
the definite odor of
cooked peppers. Should
you knock on the skin
you will hear little and
cause further collapse.

December 17 - a white
fungus has formed around
the base of the stem. the
skin in that area is very
dry (the underside is
still somewhat soft)
the odor is smoky and
metallic.

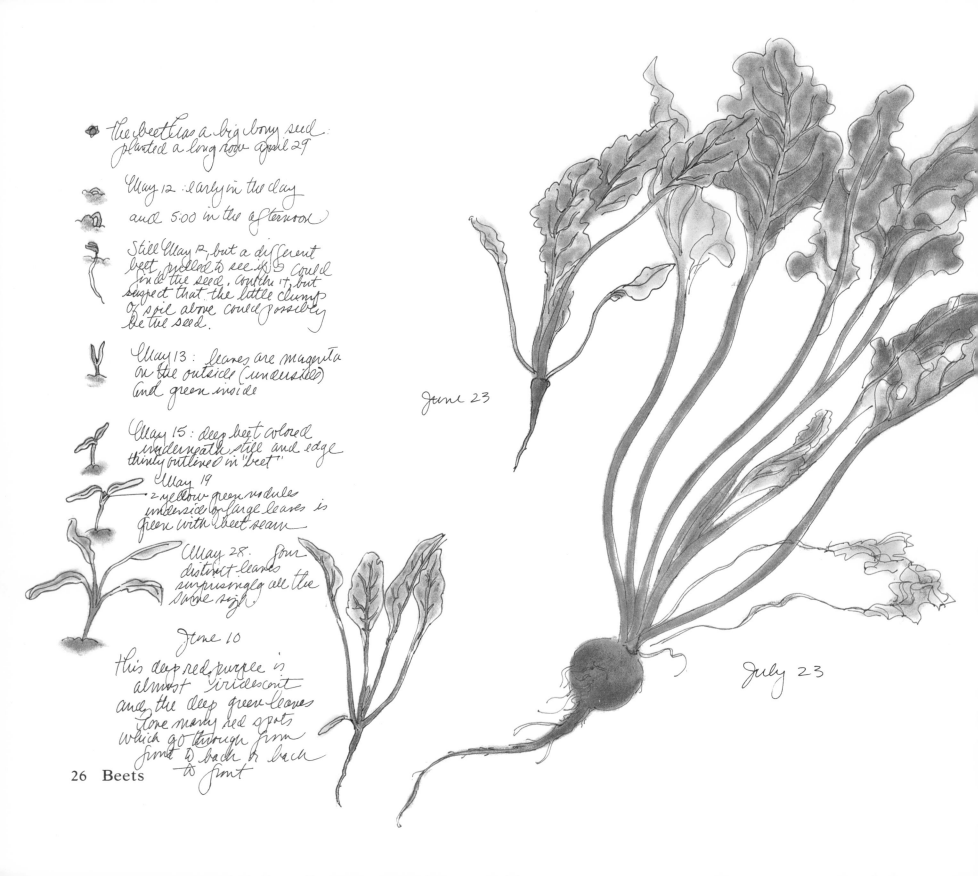

the beet has a big bony seed.
planted a long row april 29

May 12 : early in the day

and 5:00 in the afternoon

Still May 12, but a different
beet pulled to see if I could
find the seed. couldn't, but
suspect that the little clump
of soil above could possibly
be the seed.

May 13 : leaves are magenta
on the outside (undersides)
and green inside

May 15 : deep beet colored
underneath still and edge
thinly outlined in "beet"

May 19
2 yellow green nodules
underside of large leaves is
green with beet seam

May 28. four
distinct leaves
surprisingly all the
same size

June 10
this deep red purple is
almost iridescent
and the deep green leaves
have many red spots
which go through from
front to back to back
to front

June 23

July 23

26 Beets

June 20. The next year because to begin with, beets don't need the first year. This one wintered in the garden and put forth new green leaves which some small sharp teeth have already found!

July 6 and 7" tall. I think this delectable thing is going to seed; it is snatching itself in that now familiar way, but I must fashion a cage forthwith to protect it from the deer. Could I just draw one in the plot and clip for the little paws to protect his seed from his sleep?)

To spite a real cage and all myself write, the leaves have quietly grown yellow—them brown, even rather grey, but I am leaving it in the ground for yet another winter. Once in Spring, I plant the dry string thing in a flowerpot to keep it safe. Many did not recognize it as anything alive, tried it into the liquids and planted something else in the pot. How does anything ever get along? They amber leaves of grain you see, at least sing about, how do they ever manage to come to be what they start to be? It seems like one—one always comes by and lazes steps on a throws away what you're trying to grow. And so we, so people, all those little slips of paper which come into our lives, each one demanding some of our attention, each one stealing a little bit of us so that we can so rarely reading grow our amber leaves of grain. I wonder why we even try. One marvels when we can hold in our hand something we've started and finished. Mine and mine, what I admire in people is maybe not so much what they create, but just the fact that they try, they persist, they move themselves away from the fireside and the kettle and they try.

27

Zuccini seed
planted
May 20

June 10

June 11

June 12

June 15

June 17

June 21

June 30

28 Zucchini

July 17

there are tiny squashes
developing here and
there with buds
on their ends.
They haven't been
pollenated, the
flowers haven't
even opened yet!

The pale bloated
body of a blossom
has fallen from
its stem

July 22
I think the green
leaves may close a bit at night.
8:30 A.M. and the flowers are open, but
by the time I finish the drawing they have
closed. A little sun has hit them
and they have been bee-visited. I don't
know if either event was responsible
for their retreat

July 23

9:00 A.M. and closing
Someone has been eating my
leaves. Who would want
these miserable things? All
risp and rasp. Someone
with an iron throat.

(Someone with a sore throat)

On July 22 the blossoms were
closed all day since 9.30 in
the morning. I thought perhaps they'd open again in the
evening. I shall have to choose one bud and follow it
carefully throughout one full day.

31

a very tortoise-ish shell
no eyes
just Tendrils
how does he manage with
just Tendrils?

July 23
6:00 P.M.

July 24 (cloudy)
6:04 and 6:32 a small snail
is inside at the base of the flower.
Was it waiting at the entrance 15
minutes before opening time like
the ladies outside Bloomingdale's because
their trains come in from the suburbs at 9:20?

7:30   View from the
top. the snail is gone
now. If he was responsible
for pollenating the flower,
does that make him the
father or the father of the bride?

9:00 the edges of
the petals are dry-
ish and curling. &
A bee started in, then
reconsidered—Does he
know when the nectar
is gone? Does he prefer
an untouched flower?
Does he know one?

10:AM tiny hooks
at the tips of the green
central spines on the
petals.

10:50 some of
these little hooks turn
in and some out
the flower is now a
little orangeish

12:00 noon
"skin" stretched and
wrinkled. 2:00
pretty much the same,
but sunny out now.

evening
5:30
the flower does
not open again
ever

32

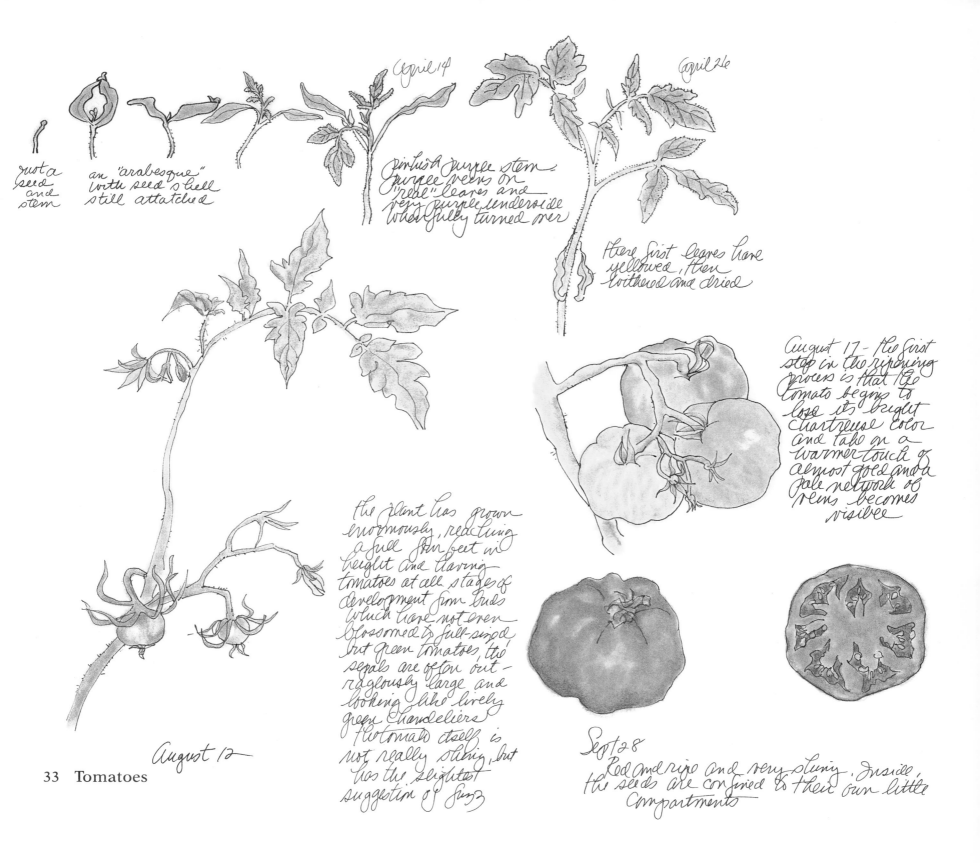

April 14

pinkish purple stem purple veins on "real" leaves and very purple underside when fully turned over

April 26

these first leaves have yellowed, then withered and dried

just a seed and stem

an "arabesque" with seed's shell still attatched

August 17 - the first step in the ripening process is that the tomato begins to lose its bright chartreuse color and take on a warmer touch of almost gold and a pale network of veins becomes visible

the plant has grown enormously, reaching a full four feet in height and having tomatoes at all stages of development from buds which have not even blossomed to full-sized but green tomatoes, the sepals are often out-ragously large and looking like lively green chandeliers. The tomato itself is not really shiny, but has the slightest suggestion of fuzz

August 12

Sept 28
Red and ripe and very shiny. Inside, the seeds are confined to their own little compartments

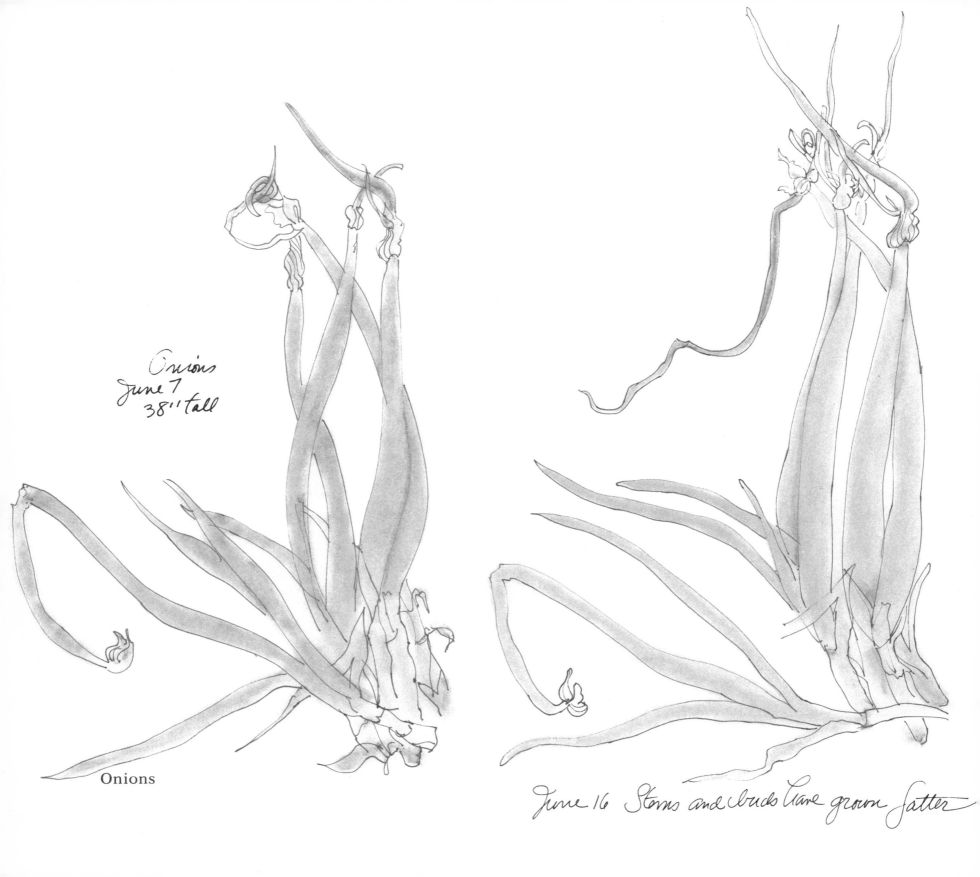

Onions
June 7
38" tall

Onions

June 16 Stems and buds have grown fatter

October 5 – the stems of
the "bulls" have fallen over
pale and dry

grey tan

35 I sense some sort of sleight of hand here. I was carefully watching the development of the top of the plant, the opening of the bulbs, their color change, the long thin shoots stretching out from them, while all this growing was going on below. There are many strong green blades where there were just a few and I don't know how and when they happened.

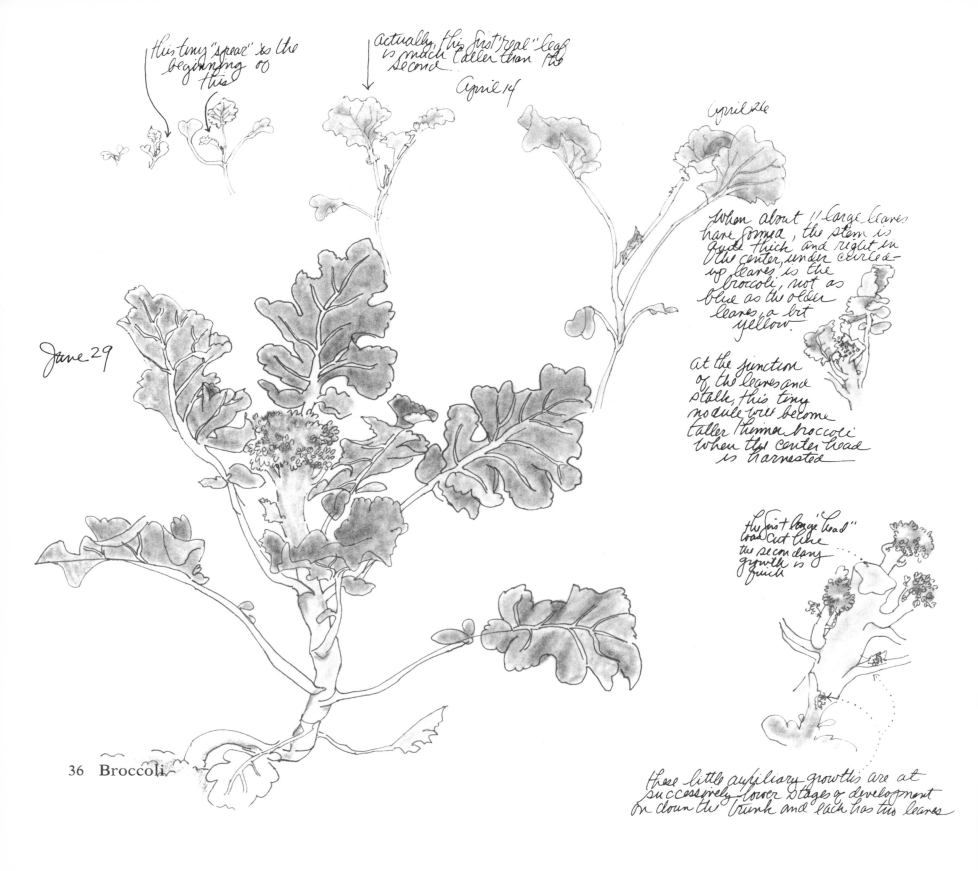

this tiny "spear" is the beginning of this

actually, this first "real" leaf is much taller than the second

April 14

April 24

When about 11 large leaves have formed, the stem is quite thick and right in the center, under curled-up leaves, is the broccoli, not as blue as the older leaves, a bit yellow.

at the junction of the leaves and stalk, this tiny nodule will become taller thinner broccoli when the center head is harvested

June 29

the first large "head" was cut here the secondary growth is quick

36 Broccoli

these little auxiliary growths are at successively lower stages of development on down the trunk and each has two leaves

a step beyond this
the tip of each branch
stretches within itself
and becomes very
airy and each
tightly closed green
bud opens up to
become a pale
yellow flower

If you allow the broccoli to
go to seed you will find that
each little "head" extends itself so that it is no longer a
tiny compact tree with a
very fat trunk. the branches
become very elongated

after the flower, the pod
and its seeds

this certainly has been a surprise. I had always fancied
broccoli growing like a veritable forest of tiny sturdy baobab
trees. the stout little trunks coming directly out of the ground
perhaps even on some good Rackham-like roots

Planted
Brussel Sprout
seeds

april 18
1:00 afternoon

april 19
1:00 with a
precise magenta
edge at top of leaf

magenta
spreading
april 20

april 21
no longer
an outline of
magenta on edge.

whole stem is
pale purply rose
and could the leaves
close together a little
at night and spread
out during the day?

April 30.
the stem is ever so
slightly greyish-
purple. The first
"real" leaf has the
sharp little scalloped
edge so like the second-
ary tooth of a child. The
child's tooth loses this
"ornament" and it becomes smooth
of edge. Will it stay with the leaf?
I am thinking of Peter Pan with
all his baby teeth and now this
seedling has had little more than
4 days of infancy, or childhood.

May 12

ah, the "tooth" has
grown and softened
and rounded. But
each new one comes in
ruffled.

May 19

June 6

July 19

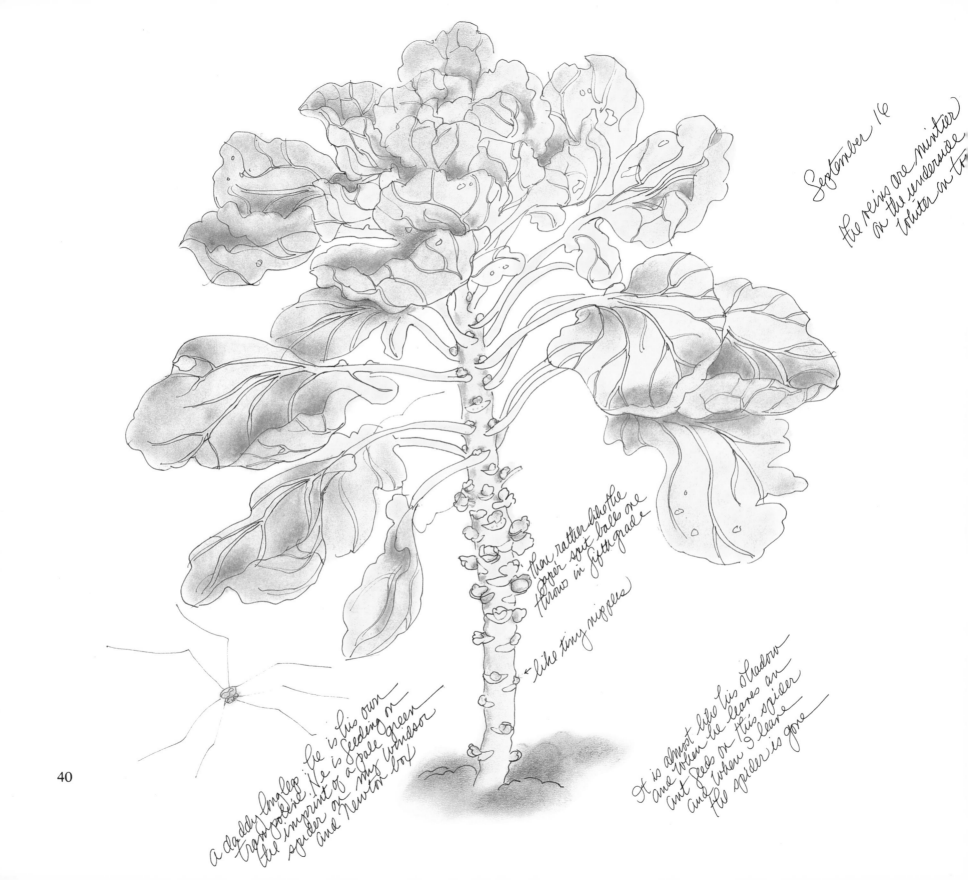

September 16

the veins are winter
on the underside
whiter on to

then, rather like the
upper spit-balls one
threw in fifth grade

←like tiny nipples

A daddy-long-leg. He is his own
trampoline. He is feeding on
the imprint of a pale green
spider on my Winsor
and Newton box

It is almost like his shadow
and when the leaves an
ant feeds on this spider
and when I leave
the spider is gone

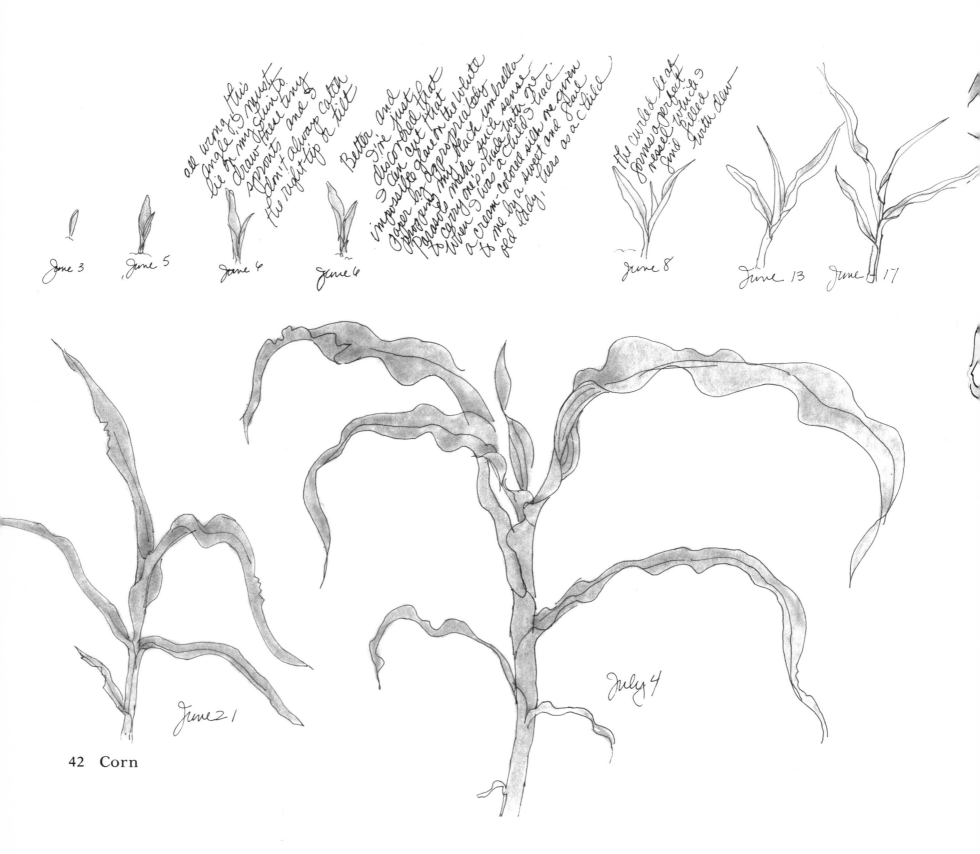

all wrong this
angle, it must
lie in my skin to
draw these tiny
sprouts, and
I can't always catch
the right tip to tilt

Better, and
I've just that
described that
I can cut that
impossible glance on the white
paper by deep appropriately
changing my hand such sense—
Parasols make such umbrella—
to carry one's shade with me
when I was a child I liked
a cream-colored silk me given
to me by a sweet and stale
old lady, lies as a child

the curled leaf
seems a perfect
vessel—drenched
and filled
with dew

June 3   June 5   June 6   June 6

June 8   June 13   June 17

June 21

July 4

42 Corn

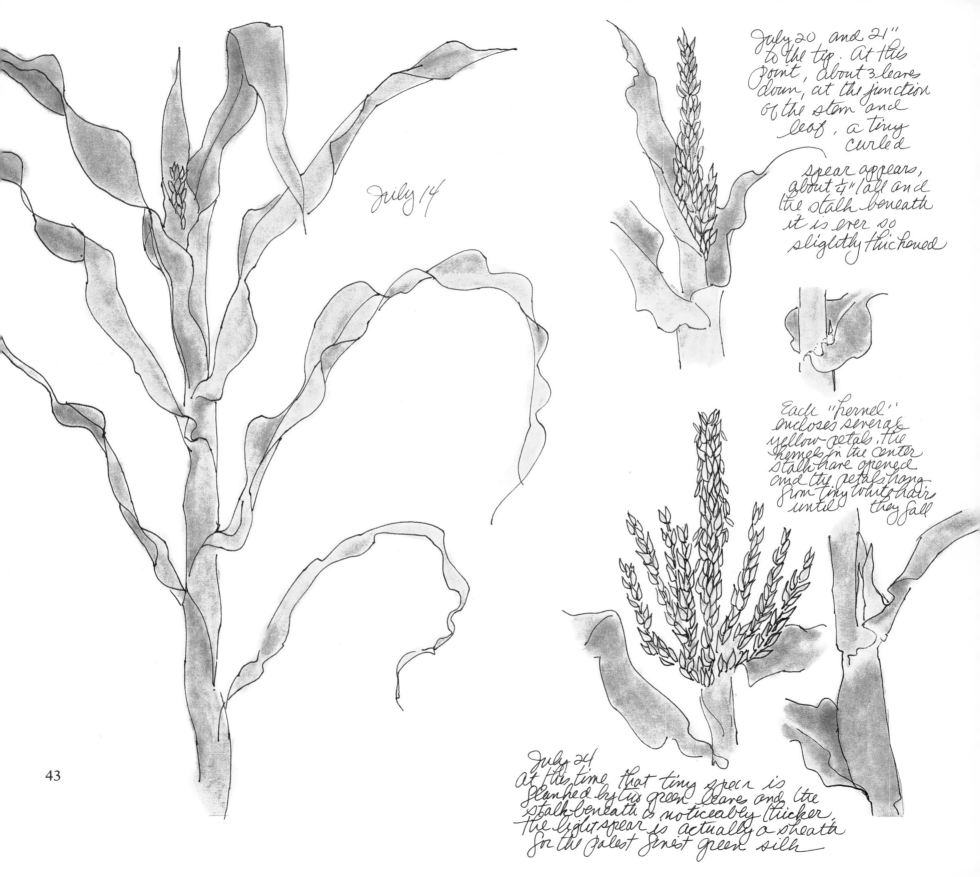

July 14

July 20 and 21"
to the tip. At this
point, about 3 leaves
down, at the junction
of the stem and
leaf, a tiny
curled

spear appears,
about ½" tall and
the stalk beneath
it is ever so
slightly thickened

Each "kernel"
encloses several
yellow petals. The
kernels in the center
stalk have opened
and the petals hang
from tiny white hair
until   they fall

July 21
at this time, that tiny spear is
flanked by two green leaves and the
stalk beneath is noticeably thicker.
the light spear is actually a sheath
for the palest finest green silk

43

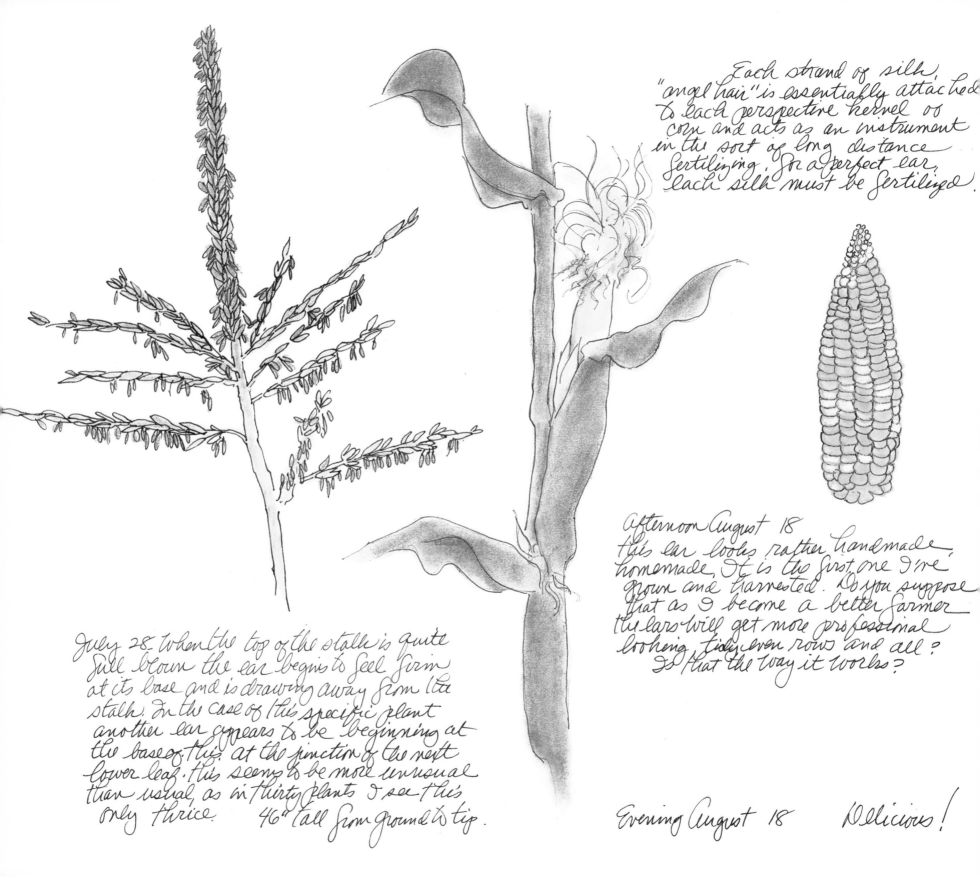

Each strand of silk, "angel hair" is essentially attached to each perspective kernel on corn and acts as an instrument in the sort of long distance fertilizing. For a perfect ear, each silk must be fertilized.

Afternoon August 18
This ear looks rather handmade, homemade. It is the first one I've grown and harvested. Do you suppose that as I become a better farmer the ears will get more professional looking, tidy even rows and all? Is that the way it works?

July 28. When the top of the stalk is quite full brown the ear begins to feel firm at its base and is drawing away from the stalk. In the case of this specific plant another ear appears to be beginning at the base of this at the junction of the next lower leaf. This seems to be more unusual than usual, as in thirty plants I see this only thrice. 46" tall from ground to tip.

Evening August 18   Delicious!

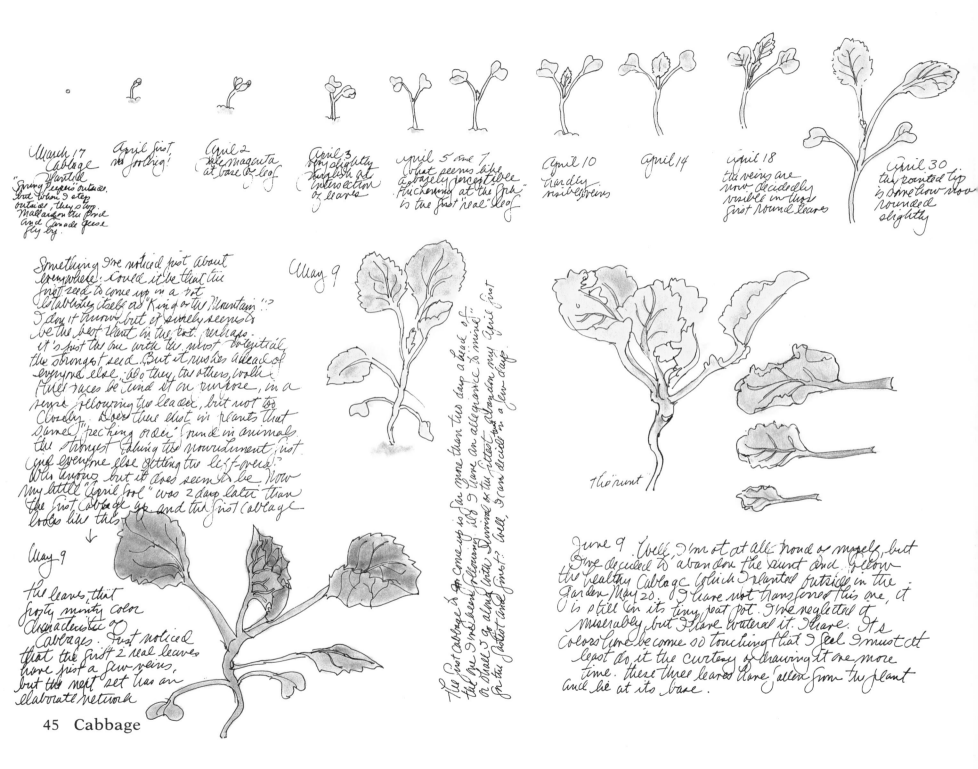

March 17
Cabbage
Planted
"Spring Peepers' outside,
but when I step
outside, they stop.
Mallards on the pond
and Canada geese
fly by.

April first,
no fooling!

April 2
like magenta
at base of leaf

April 3
very slightly
purplish at
intersection
of leaves

April 5 and 7
what seems like
a barely perceptible
thickening at the top
is the first "real" leaf

April 10
hardly
visible veins

April 14

April 18
the veins are
now decidedly
visible in those
first round leaves

April 30
the pointed tip
is somehow now
rounded
slightly

Something I've noticed just about
everywhere: could it be that the
first seed to come up in a pot
establishes itself as "King of the Mountain"?
I don't know, but it surely seems to
be the best plant in the pot. perhaps
it's just the one with the most potential,
the strongest seed. But it rushes ahead of
everyone else; do they, the others, with
their races behind it on purpose, in a
sense following the leader, but not too
closely. Does there exist in plants that
same "pecking order" found in animals,
the strongest taking the nourishment first
and everyone else getting the leftovers?
Who knows, but it does seem to be. Now
my little "April Fool" was 2 days later than
the first Cabbage up. and the first Cabbage
looks like this
↓
May 9

the leaves, that
frosty minty color
characteristic of
Cabbages. Just noticed
that the first 2 real leaves
have just a few veins,
but the next set has an
elaborate network

May 9

The runt

The first Cabbage to come up is far gone than the
the one I'm about following. did I have an allegiance to mind? Or shall I go along with Darwin or the fittest and abandon my April first for the fastest and first? Well, I can decide in a few days.

June 9. Well, I'm not at all proud of myself, but
I've decided to abandon the runt and follow
the healthy Cabbage which I planted outside in the
Garden May 20. I have not transferred this one, it
is still in its tiny peat pot. I've neglected it
miserably, but I have watered it. I have. Its
colors have become so touching that I feel I must at
least do it the curtesy of drawing it one more
time. these three leaves have fallen from the plant
and lie at its base.

45 Cabbage

June 9th (Johnny's birthday) and this is
that first spring cabbage

46

This most voluptuous of vegetables with all its varying greens, olive, mint, lime, sage, Saratoga, somehow reigns in the garden. Do all the other vegetables stand in its presence? It is ornamental garden harvest. I have known people to leave it in the garden because it looks so nice there although it would go well with dinner. I have known myself to leave it in the garden because it looks so nice there although it would go well with the dinner.

It seems that the peasant cooks it for his soup the rich never touch it, but fashion elegant ceramic tureens in its shape and everyone in between makes cole slaw.

On this tall cabbage, one leaf had an especially beautiful curve in its veining. I began to wonder why this specific curve was so much more beautiful to me than the leaf's other curves. Peter Stevens tells us in *Patterns in Nature* that a tree's branching resembles that of rivers and human arteries. This throws me for an absolute loop! Could it be that we are drawn to certain shapes simply because the same shapes are within us? Could the curve of the cabbage leaf vein be precisely the curve as a vein in my own heart or on my eyelid?

Has man chosen and repeated and loved the arched doorway, not because it is beautiful in itself, but because it is so sympathetic to the proportions of the human figure? With rounded head in rounded archway we feel very much at home. Would the newt then love what is long and low?

I've often wondered why little scenes in nature seem perfectly balanced, the rocks on the riverbank just right in size and interval. A group of cows on a hill look like they've arranged themselves as one would arrange them for a painting, but they are merely exercising their natural social distance. The closest I can come to putting this into a sentence is: We judge nature by the standards she herself sets. We find something beautiful not only because it is what it is, but also because it does what it does. Perhaps that's what Keats's phrase, "Beauty is truth, truth beauty" has meant all this time without my knowing it.

black dots

This red dragonfly
tarries here on the paper
seductively folding
and unfolding
its wings
flirting
Sept. 8.   and
preening

and it seems to me, asking,
"take my picture"
"take my picture"
and so, I do.

I was struck by this plant,
so seemingly in the dramatic
throes of something important:
Its stance, its contraposto suggest
such effort and determination.
I think these numerous appendages
are preparing to branch and flower.
It seems that cabbage does not seed
the first year. None of mine have although left long in the ground. This
plant was in the garden of a friend
and apparently passed by at harvest
time last year. It looks very
brussel sproutish with all its little
arms, but I have been assured that
it is quite cabbage.

49

These wonderful
peachy mauve colors
developed when the leaf
parted the plant. I think they
are its swan song. I have
kept the leaf. It is
smaller now, dry and
curled. It sits on my
window ledge. It looks
like an ocean wave. Light
50 shine through its little holes

THE FLOWER GARDEN

My first memory of flowers is of those from the cemetery. So many times we played hide-and-go-seek there and some of those times I'd take flowers from the graves to put in my room. They were not pretty. Their short wired stems made them staunch and their faded ribbons made me sad. I wonder now why I brought them home at all.

The lilacs were the real flowers. The tallest bush in the world grew in the yard next to the cemetery and its tip-tops taunted us from above the fence. They swayed gloriously, they smelled like heaven, and I'd stand on my brother's shoulders to pick them. Their owner sat at his window all days and all nights monitoring the motion of his bush. When an uncharacteristic jiggle suggested small thieves, he'd leap from his post and roar toward his plant. Not lion, not lynx frightened us more. We fled on jumbling legs, with fear of jail, with pounding chests, and sometimes, with lilacs.

I did not grow up having a garden. Our dry gray house stood straight up from the ground without a bit of bush to soften its transition from the earth. My brother planted grass and a short string fence. We fluttered the fence with strips of white rag, but the tall tree and its long swing kept the grass from growing. And yet—in the far corner of the yard,—grew a small bush of pale roses.

The wonderful thing was that our roses bloomed in time for the Memorial Day parade. Behind the band, behind the children on bicycles decorated with red, white, and blue striped crepe paper I followed, proud to have flowers to carry for the soldiers.

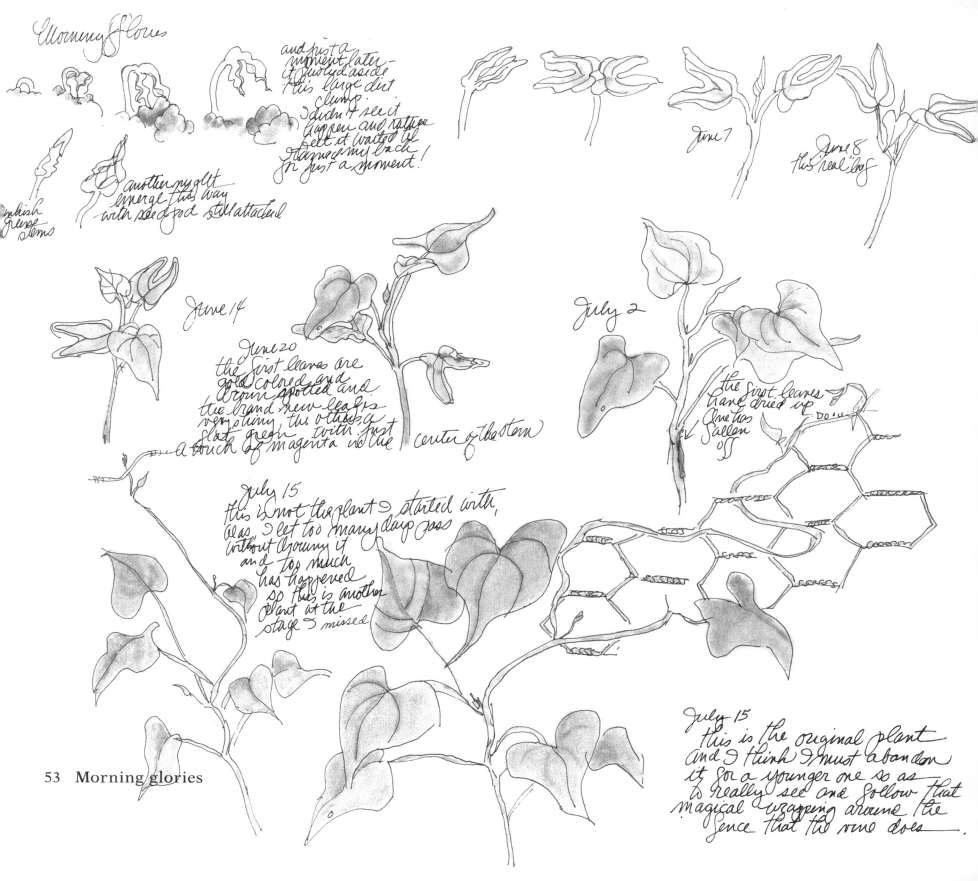

Morning glories

reddish grease stems

and just a moment later it quietly aside this large dirt clump. I didn't see it happen and rather felt it waited till I turned my back in just a moment!

another my glot emerge this way with seed pod still attached

June 7

June 8 this "real" leaf

June 14

June 20
the first leaves are gold colored and brown spotted and the brand new leaf is very shiny, the others a flat green with just a touch of magenta in the center of the stem

July 2

the first leaves have dried up one has fallen off

July 15
this is not the plant I started with, alas, I let too many days pass without drawing it and too much has happened so this is another plant at the stage I missed

July 15
this is the original plant and I think I must abandon it for a younger one so as to really see and follow that magical wrapping around the fence that the vine does.

53   Morning glories

July 17

two days ago
the tip was just
beginning to
start reacting
around like
a vine

July 19

After a good
rain. I've had
to compress the
fence a bit to get
the top of the vine
in. One thing I've
just learned is never
take anything for
granted. I thought
the lower section of
the vine would stay the
same and just the
end change, but instead
it has tightened up in
spots, shifted a bit
and begun to branch
out. The rusty color
on the stem has
nearly disappeared
completely except for
the lowest point
where it is close to
the ground.

Somehow there
seems to be no
way of telling when
the first flowers

will appear. Another
plant nearby already
has had two blossoms
& many buds. It does
get the sun a bit earlier,
but then there's another vine
on the sunny side of that
one which hasn't
bloomed yet either.

flower
bud

end

bud

It is long long later now. I am
looking back at these drawings
and I know now that the morning
glory vine moves only in a clockwise
direction (some vines move counter clockwise).
Its motion had seemed rather willy
nilly to me at the time. I have often
wanted to help the vine find the fence
and so interfered, but not knowing
the pattern I have from time to time twisted
a vine the wrong way around the fence
(very unscientific of me). Unless I have
practically tied it into position it
unwinds itself, falls down and does the
thing in its own way and its own time.

The English Michael Flanders
and Donald Swann have a little
song called Misalliance about the
honeysuckle and the bindweed
(wild morning glory) which fall
in love.

But alas, one twines to the
left and one twines to the right...

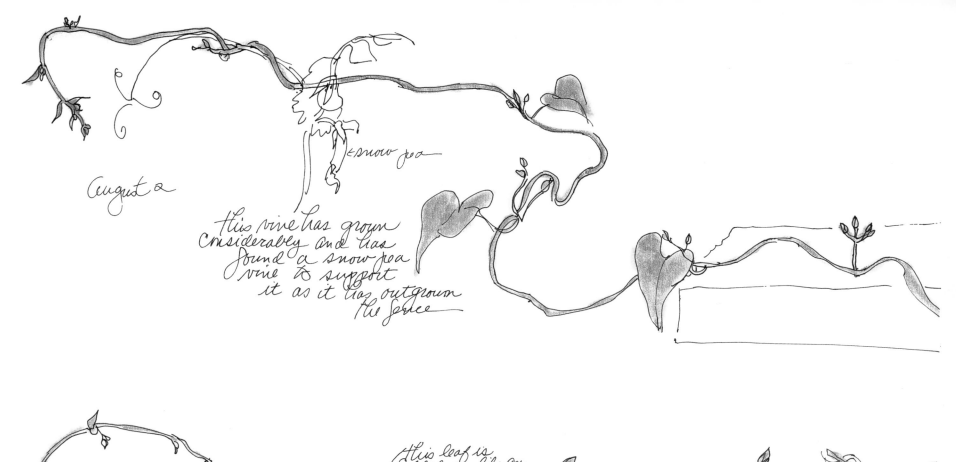

August 2

← snow pea

This vine has grown
considerabley and has
found a snow pea
vine to support
it as it has outgrown
the fence

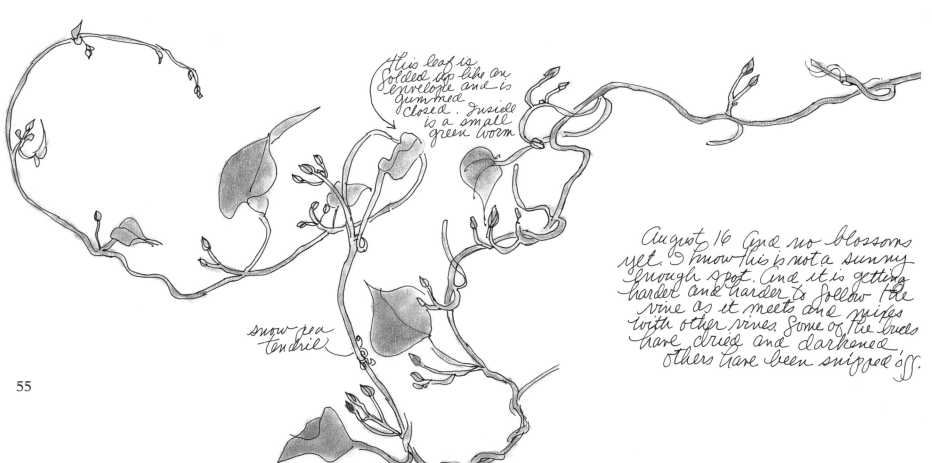

This leaf is
folded up like an
envelope and is
gummed
closed. Inside
is a small
green worm

snow pea
tendril

August 16 And no blossoms
yet. I know this is not a sunny
enough spot. And it is getting
harder and harder to follow the
vine as it meets and mixes
with other vines. Some of the buds
have dried and darkened,
others have been snipped off.

September 4

Nipped in the bud —
What else could I say?

the bud develops

the seed develops

56

MORNING GLORIES

There have been flowers in my life since my hide-and-go-seeking, honeysuckle-sucking, snapdragon-snapping days. There were times when a young man would bring me a single red rose. But it wasn't until I saw morning glories that I felt so exhilarated by a flower as to inquire into its being and behavior.

There were so many things I wanted to know about morning glories. I courted them, I suppose. I wanted to know when they got up and how quickly they opened. So I would often go out all the long night and early morning with my flashlight. I would see the day and the bud begin to open at the same time. The bud is purple-pink in the purple-pink of the early morning, slips into violet, and flower as the sky does. It becomes midday-blue at midday, then retraces its colors in closing as does the day.

I discovered that the vine always spirals in the direction opposite from the furl of the bud. (I wonder if this keeps the plant from getting dizzy?) I've watched a vine grow an alarming three inches in ten hours. As I drew the drama of these exuberant blue trumpets, I felt the poignancy of their being so alive and so short-lived. They dare not nap. Dare any of us?

The more I drew morning glories, the more I felt I was drawing more than morning glories.

6:15 A.M. the tip is very blue

7:00 A.M. the white stripes have a very narrow edge ridge which is now slightly suggesting red-violet at the lower tips and is paler at the base. the base green at the base is no longer green but yellow.

Morning glory at noon

6:00 P.M. the changes from 3:00 to 6:00 are not being so dramatic so Three hours span time, but the color is decidedly red violet

The stripes are blue violet and pale pale lavender the lavender will become the five pointed star in the flower

8:30 A.M.

the day before the blossom at 3:00 A.M. like this at 4:45 A.M. the tip is red. violet →

7:30 A.M. As the flower opens it becomes apparent that the center of that deep fold is the narrow white line between the definite white star spokes

8:00 A.M. the blue is really blue now having lost the violet feeling, and when you look inside the flower the base is buttery yellow

2:45 A.M. this is unusually early I think for the flower to begin to close, but it was a very windy day. at first the wind made the edges flap inward just a bit and the flower seemed encouraged by this suggestion of closing to continue on its own and with such rapidity that I simply could not draw fast enough. Between 2:45 and 3:00 the edges were definitely curled inward and under this shy blue was rapidly changing to blue violet and the ridged edge were slightly red violet as they were at 7:00 A.M.

7:30 P.M. Everything is tidily tucked in and under the flower will not open again. It has had its "moment in the sun."

58

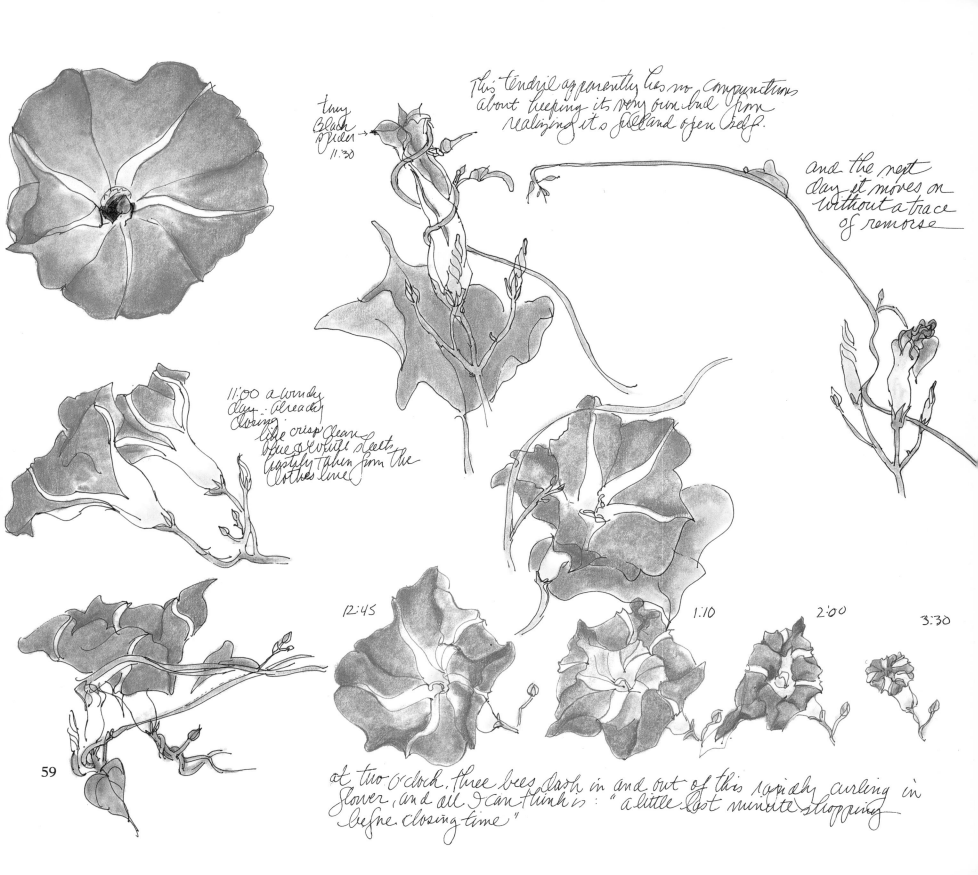

This tendril apparently has no compunctions about keeping its very own bud from realizing its full and open self.

tiny Black Spider →
11:30

and the next day it moves on without a trace of remorse

11:00 a windy day. Already closing. like crisp clean blue & white sheets hastily taken from the clothes line.

12:45   1:10   2:00   3:30

59

at two o'clock, three bees dash in and out of this rapidly curling in flower, and all I can think is: "a little last minute shopping before closing time"

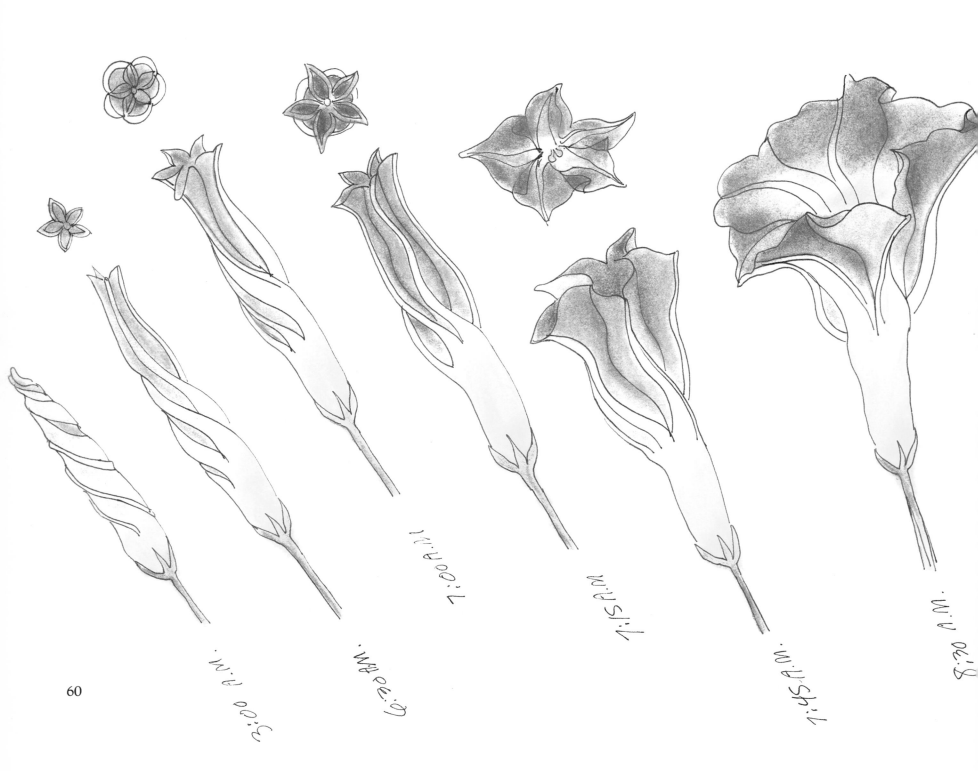

3:00 A.M.

6:30 A.M.

7:00 A.M.

7:15 A.M.

7:45 A.M.

8:30 A.M.

12:00

2:30 P.M

2:45 P.M.

3:30 P.M.

4:20 P.M.

the next day

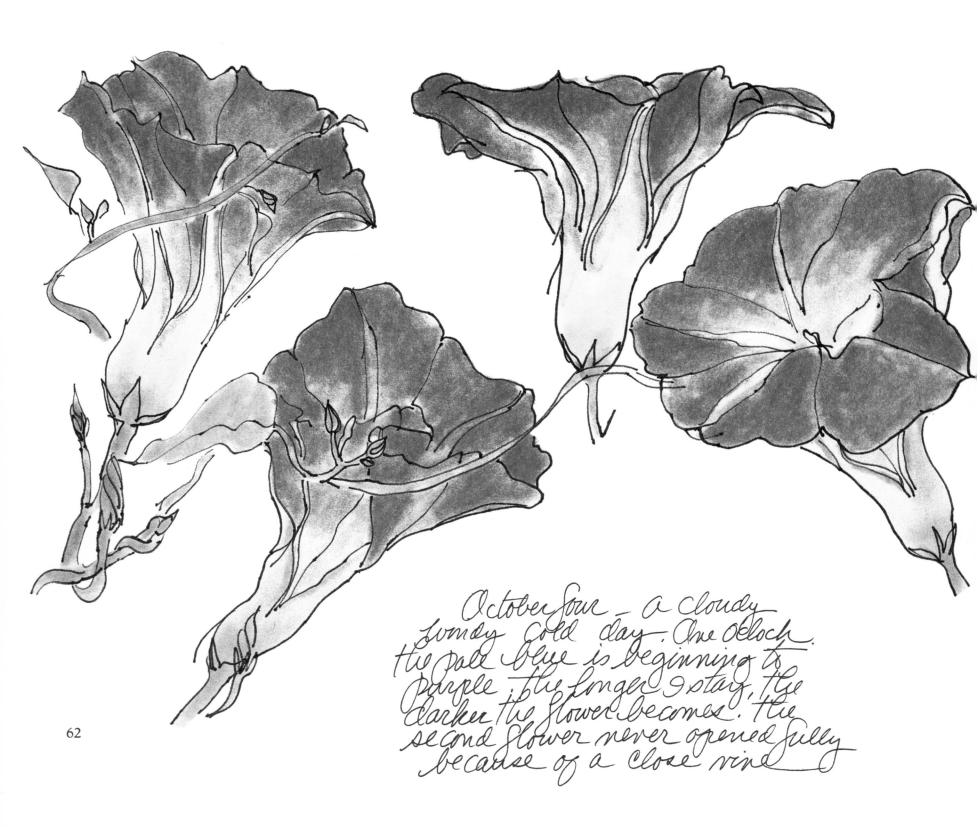

October four — a cloudy
Sunday cold day. One o'clock.
the pale blue is beginning to
purple. the longer I stay, the
darker the flower becomes. the
second flower never opened fully
because of a close vine

62

October five

The proud blossoms of yesterday

Every flower in the field and garden, on roadside or rivers edge, takes it's notice-me stance, spreads out and smooths down its petals, posts a little here-I-am-fragrance to every passer-by in the hopes of attracting a pollenator of sorts.

And, so the young girls, all arched and alert in their summer dresses turn around quickly so as to make a little billow to their skirts, to spread out that colorful surface to the fullest, put on a little scent for the passer-by and hope to attract a pollenator of sorts.

This drawing looks like something someone would paint on a tray. I don't like it at all. Is it because of the flamboyant full flower? Why is it that the full flower often seems a little tawdry? Now that is an unfair word and the wrong one, I'm sure, but the lush and prime is so much less interesting than the forming or the failing flower. It stands still and doesn't touch us in the same way.

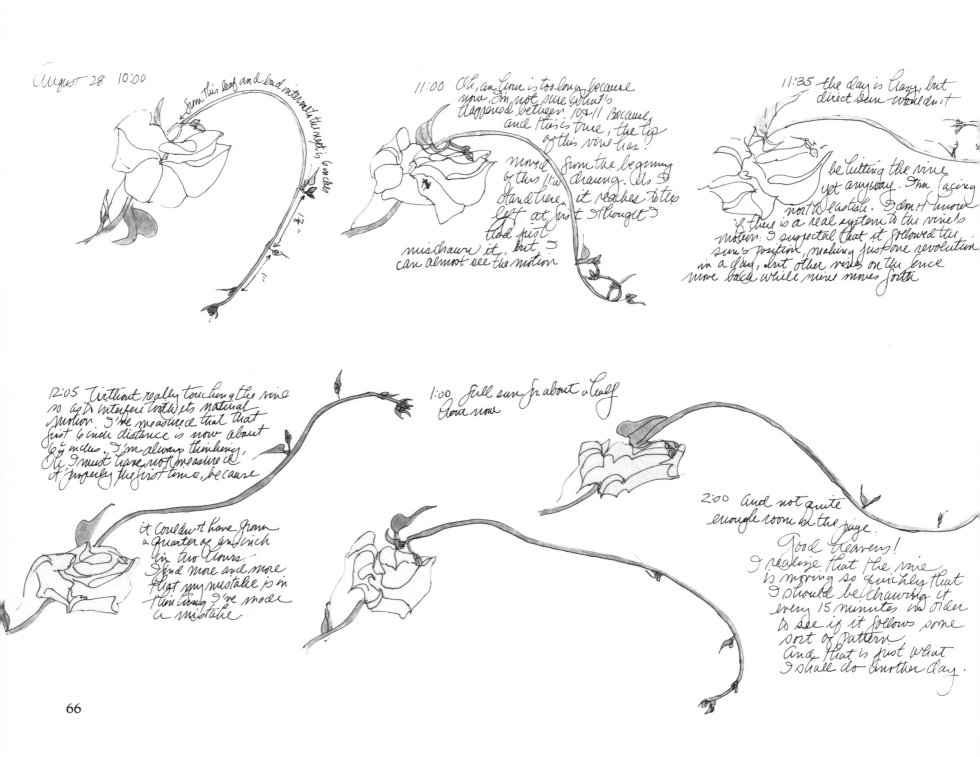

August 28 10:00

From this leaf and bud intermittently tip is 6 inches

11:00 Oh, an hour is too long, because now, I'm not sure what's happened between 10 & 11. Because, and this is true, the tip of this vine has!

moved from the beginning of this 11:00 drawing. As I stand here, it reaches to the left. At first I thought I had just

misdrawn it, but I can almost see the motion

11:35 the day is hazy, but direct sun wouldn't

I'll be cutting the vine yet anyway. I'm facing northeastish. I don't know if there is a real system to the vine's motion. I suspected that it followed the sun's position, making just one revolution in a day, but other vines on the fence move back while mine moves forth

12:05 Without really touching the vine so as to interfere with its natural motion, I've measured that that first 6 inch distance is now about 6½ inches. I'm always thinking, Oh I must have not measured it properly the first time, because

it couldn't have grown a quarter of an inch in two hours. I find more and more that my mistake is in thinking I've made a mistake

1:00 full sun for about a half hour now

2:00 And not quite enough room on the page.
Good heavens!
I realize that the vine is moving so quickly that I should be drawing it every 15 minutes in order to see if it follows some sort of pattern.
And that is just what I shall do another day.

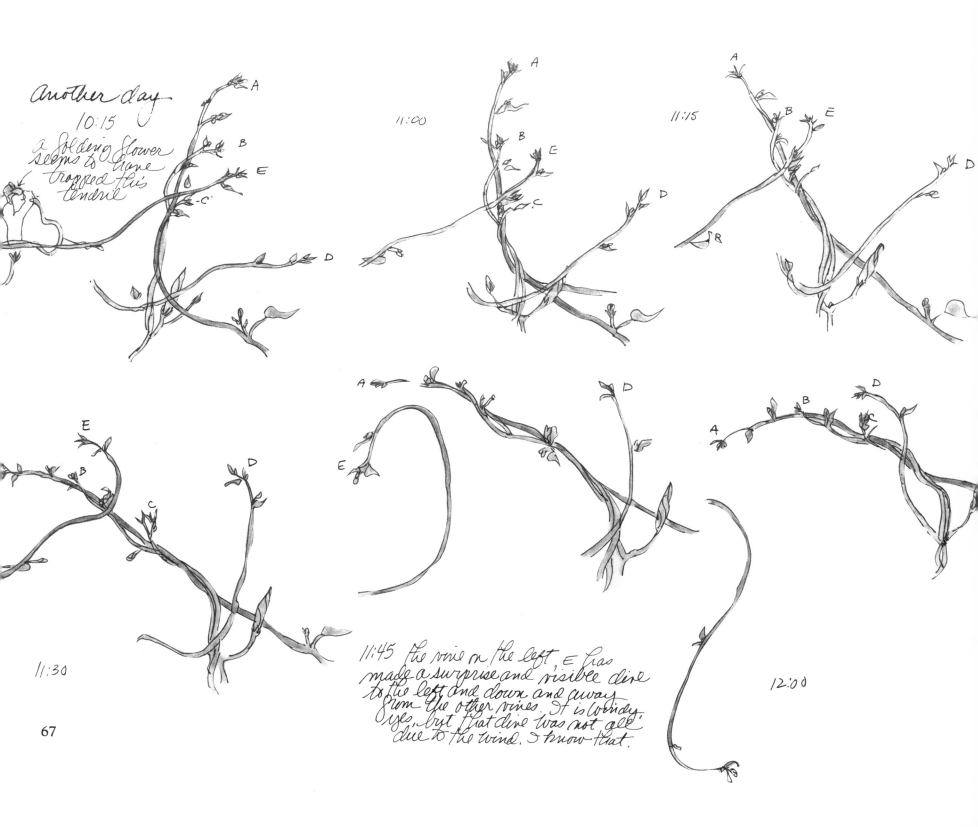

Another day
10:15
A golden flower
seems to have
trapped this
tendril

11:00

11:15

11:30

11:45 The vine on the left, E has
made a surprise and visible dive
to the left and down and away
from the other vines. It is windy,
yes, but that dive was not all
due to the wind. I know that.

12:00

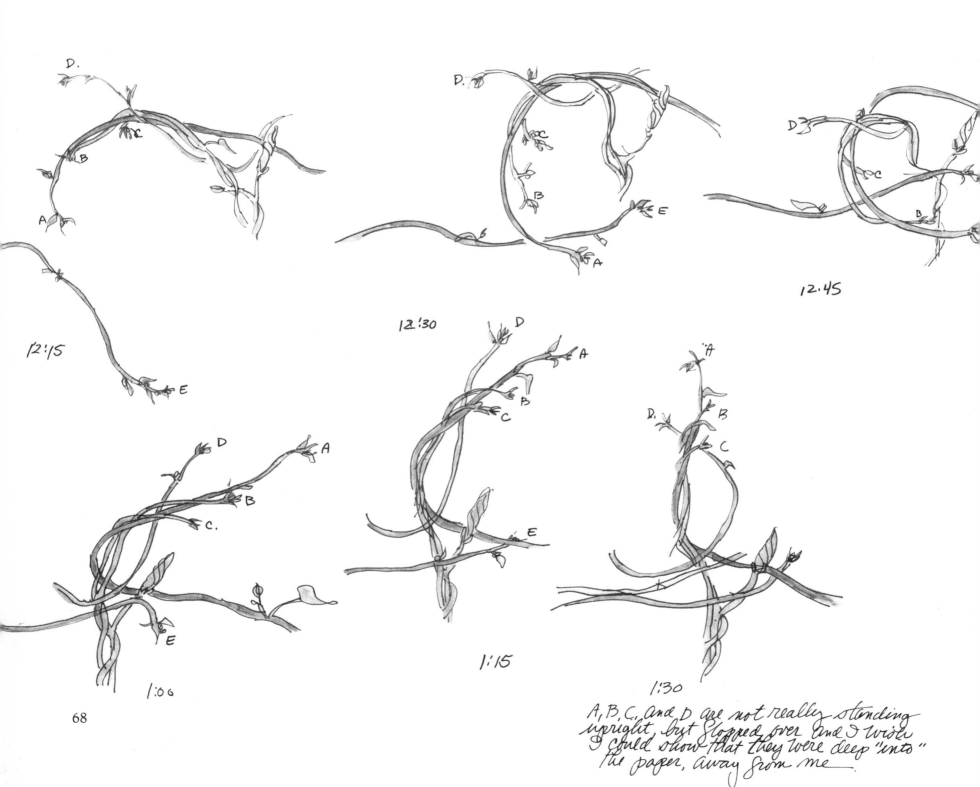

D.

B

A

12:15

E

12:30

D.

C

B

E

A

12:45

D

C

B

D

A

B

C.

1:00

E

D

A

B

C

E

1:15

"A

B

D.

C

1:30

A, B, C, and D are not really standing
upright, but flopped over and I wish
I could show that they were deep "into"
the paper, away from me.

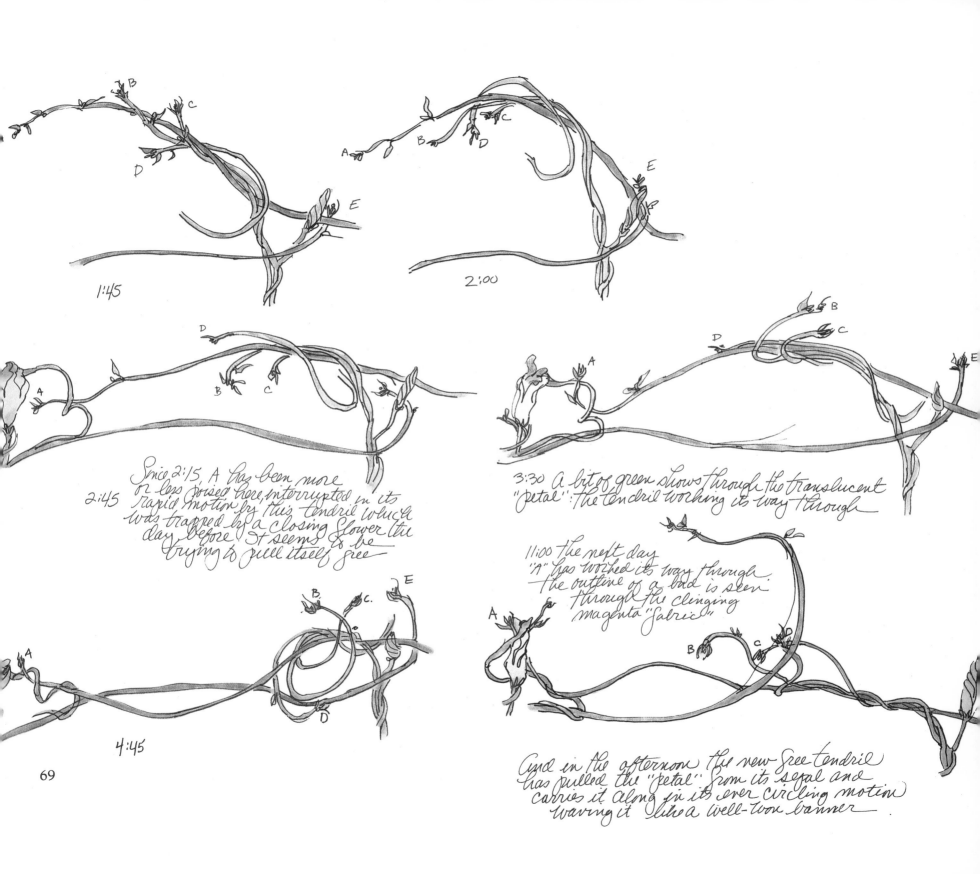

1:45

2:00

Since 2:15, A has been more
or less poised here, interrupted in its
2:45 rapid motion by this tendril which
was trapped by a closing flower the
day before. It seems to be
trying to pull itself free

3:30 A bit of green shows through the translucent
"petal": the tendril working its way through

11:00 the next day
"A" has worked its way through.
the outline of a bud is seen
through the clinging
magenta "fabric"

4:45

And in the afternoon the new free tendril
has pulled the "petal" from its sepal and
carries it along in its ever circling motion
waving it like a well-worn banner.

69

Monday September 25 — I want to follow for a few days the events of
this small square of morning glories on my chicken wire fence. I cannot
believe the multitude of the buds and the tinyness of some of them.

Tuesday
September 26
Yesterday's flowers are folded now and the tendril has grabbed another
tendril nearby and they are entwined together now (upper right near the
faded, dusty jimhish dead flower, but dead is an awful word isn't it?

Wednesday
September 27   It is quite cold so that some of the buds
which were ready to unfurl cannot open all the way. They need more
heat. And as the afternoon progresses the ones which did open fully
because they had started to open the day before didn't close. It is taking
two days to do what they ordinarily do in one.

Thursday   September 28
the vine circles all day long! On Sunday, three days later I don't even
recognize the spot. So many plots are going on here, such drama upon
drama and this is just one foot square on my fence. When I think of
the whole fence and what is beyond it, under and above it, I am dizzy.
It is frightening and wonderful.

DAY LILIES

I've been told that all parts of the day lily are edible. I haven't tried any parts yet, but I suddenly remember a flower incident of several years ago. I had been invited to an extraordinarily elegant dinner party. The table was set with a grand display of silverware, much lavish Swiss embroidery. A servant attended the meal and did everything for us except cut our meat. After the entrée each guest received one very large, red, spikey, outrageous, exotic flower on a plate. All I could think was, Dear God in heaven, how do I eat the thing? No one seemed concerned with the problem at hand and no one seemed to eat it either, so I blithely ignored it. Shortly thereafter the servant took it away. I guess it was intended to be an intermission, an Oriental sort of interval, but it was enough to scare a poor country girl to death.

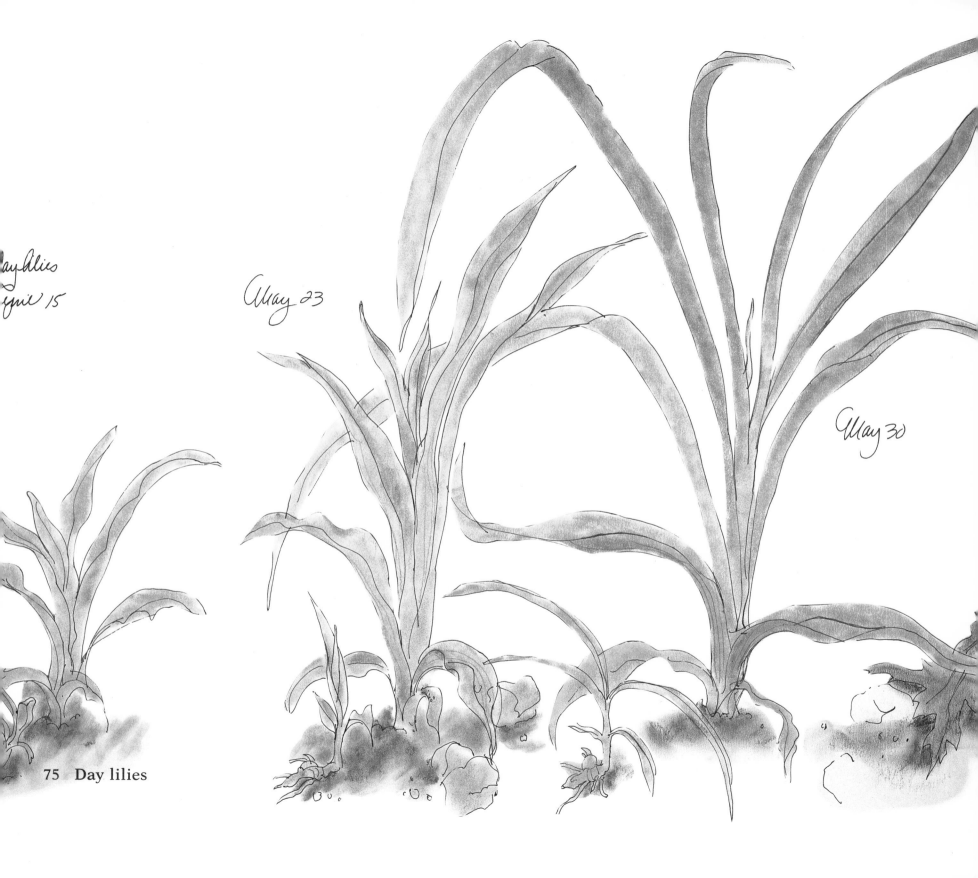

Day lilies
June 15

May 23

May 30

75 Day lilies

June 15

bud slightly
yellow

June 23

July 3
rain &
retreat

July 9
noon

77

July 19 - 2:00 P.M.
Chandeliers opening slightly

July 20   5:17 A.M.

6:10 A.M. The tips of the
petals seem to have the
tiniest possible white

Looks, keeping the
and tightly closed until
the right time and
making natures own
"tick-tock-tick-tock"

6.58 A.M. The green
tips are lighter but
barely perceptibly so

78

Thunder and a straight up and down waterfall

Rain again. the shower and I and the drawing I made under the umbrella I tried to fit I balanced as my head and one shoulder and

5:00 in the evening

the day after the party the crepe-paper decorations hang

damp and twisted

79

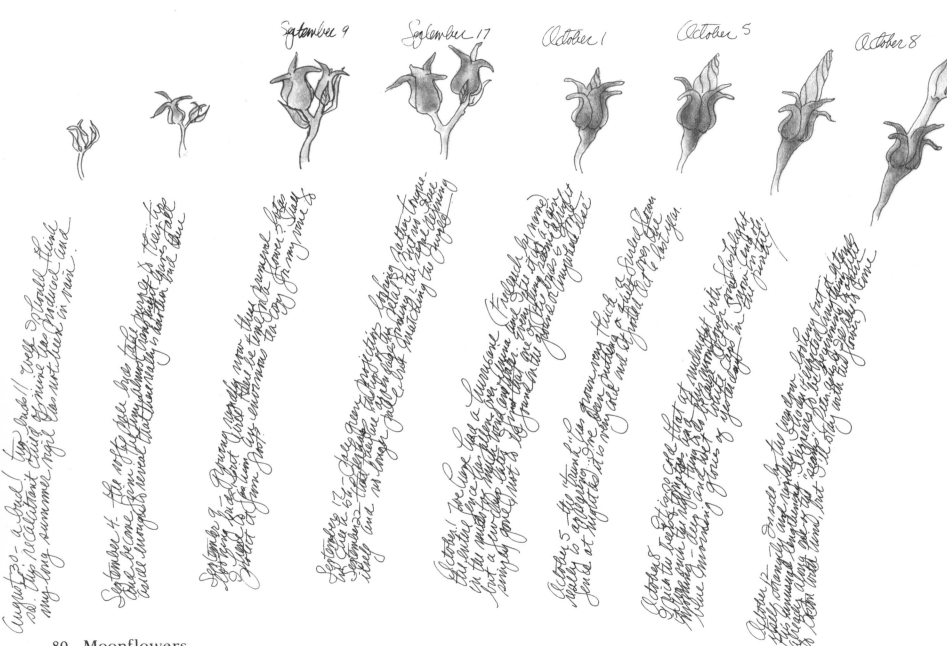

12:15 P.M. The excitement and anticipation are in the air on October 13. Everyone of the bud and my stalk — everyone is measuring the bud — everyone else is watching with the wild guessing of when the flower will open — and the cast off.

3:30 A.M. October 14. I look for a bit of my own. The day is but I find more by moonlight. I tried and seem fitting. By moonlight it is evident in a sort of midrange. First flower in full spread of a sort of midrange.

5:00 A.M. The tightly closed bud slightly and slowly beginning to begin to open ever so definitely leaning toward the light. Continues open but the change. The unfolding. 8:00 does as does 9:00 — 10:00 nothing same cream filled which to has stopped. It is the find from an hour ago.

October 15 — Alas. The flower did grow any and then collapsed. I am all flopped. It hung limp and there. The sepal edges clearly deflowered. The not green so it because I pulled every air and froze it but it cannot be ever... the blooming and frost of night because of imagining me? magenta... it a drew it. Because of the starts of closing a June. Oct. 14? Because I didn't know how closing a June. It can't be open and it needed that anyway it. Cause enough. Did I care to much over... too things to death? So it it feel it enough a day. The possibility of the cut summer. Once star I... Everyone forgets a trying summer... following all time preliminary of age. The day growing ending — especially the I take me down the garden path. I did not intend it because I see what a more flower simply. Turning pages — and I must wait until next summer.

noon

September 23
4:30 P.M.

4:45

4:50

5:00

5:10

5:20

6:00

The moonflower puts things in perspective better than anything else I know.

4:30 A.M.
September 24

12:00 noon

12:30

1:00 P.M.

1:20

6:15 P.M.

6:00 P.M.

the next day
and slightly pinkish

It is its fairest, fullest, and most fragrant while we are all asleep

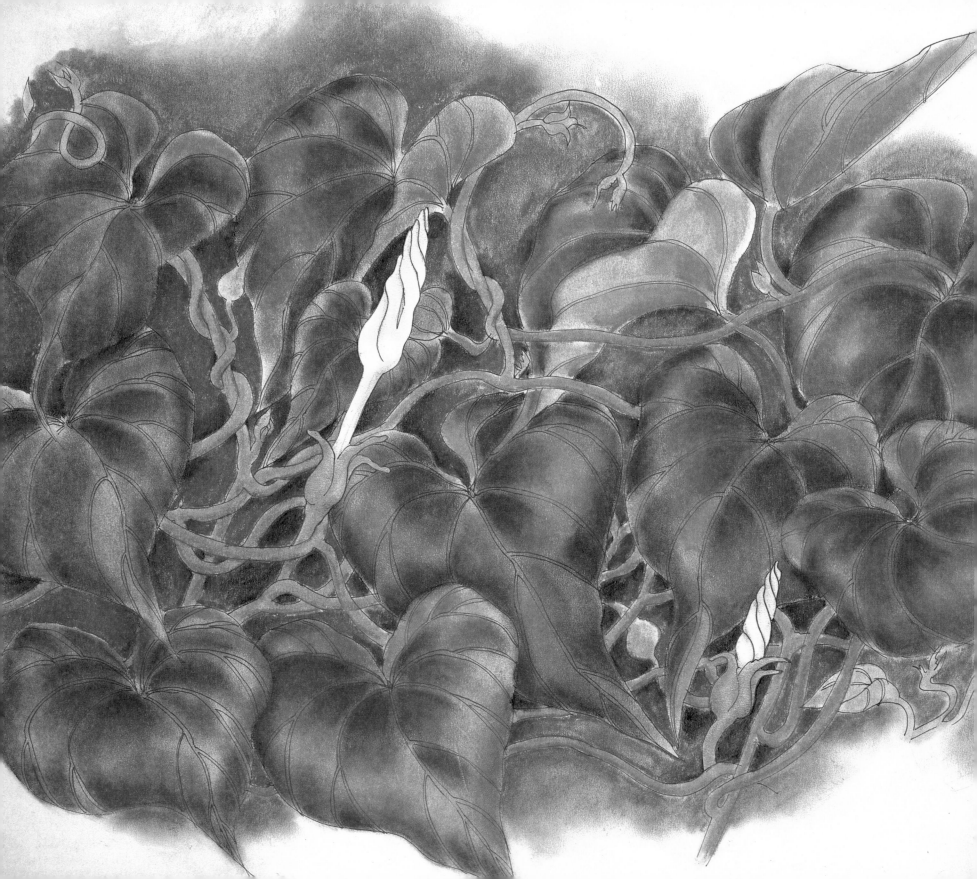

Baudelaire refers to the beauty of a woman past her prime. Lately I am drawn to things past their prime. Ten years have passed since I began these studies. My children have gone from growing to grown. Supple has changed to subtle, lilt become tilt. Those flowering, fruiting years are slipping toward the frailing years. I look at things differently now. I look at different things now.

I look at the dying flower. It dries and becomes exquisitely etched and furled. A flower plays its hand of colors, discards the needless ones, carefully selects the elite, and goes out of the game with these subtle few.

I have kept many fragile spent blossoms, their shadows stronger than their selves. As I sat and drew the weightless translucent rose leaves I realized why I ever drew anything. It's the only way I know to say "mm-hmm!" Someone else may look at the sunset, say "mm-hmm," and let it go at that. I don't dismiss this exclamation as any less an appreciation than my pulling out pastels and trying to put sunset on paper. My exclamation just takes longer.

Perhaps I should pastel my body as sunset, paint my limbs as flowering vines, and get this wish to be nature out of my system. Then maybe I could relax.

the bones of last summer this spring
May 14, winter bleached:
the moonflower vine

87   Moonflower vine

Three stems of the Fairy Rose

88   Fairy rose

Three Amaryllis Flowers

89 Amaryllis

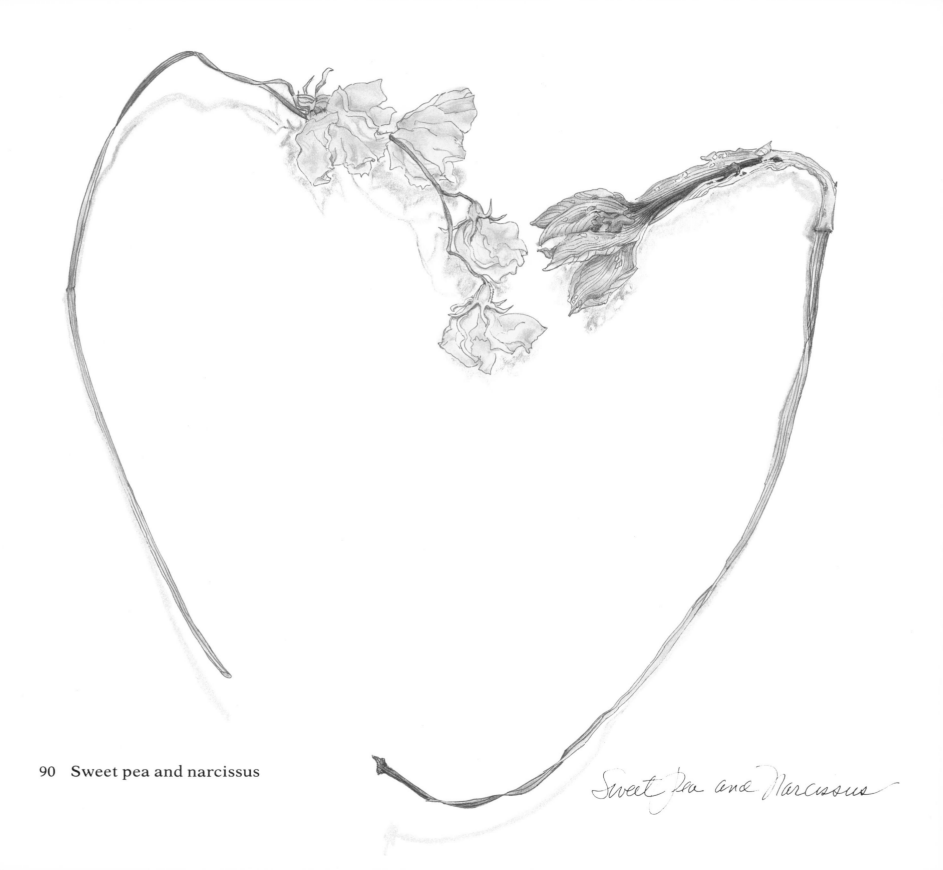

90 Sweet pea and narcissus

Sweet pea and Narcissus

THE ANIMAL GARDEN

Whenever I draw plants, there are animals. The kingdoms never separate.

A dragonfly primps and preens on the edge of my drawing, seeming to say, "Take my picture, take my picture." A daddy longlegs bounces on the trampoline of himself and feeds on the squashed imprint of a pale spider on my Windsor & Newton box. When he leaves an ant feeds upon it; when the ant leaves the imprint is gone. A squirrel, with tail clung close to his head like a parasol against the mosquitoes, seems unconcerned by my presence. He finds a large toadstool and eats it like a chocolate-covered ice cream bar—the thin layer of brown on the outside first, then the white inside, then the stick.

When quietly working, I might hear such commotion as to quite alarm me, a vast rustling of leaves. I look in all directions for the boy scout jamboree it surely must be. It is a bird. It is always a bird.

There is the neighbor's cat rubbing against my legs and knocking me over, making it harder still to gain secure footing and draw on the steep day lily hill. I keep chasing him away and having him come back and push me over again, chasing him away and having him come back and push me over again....

93 Birds

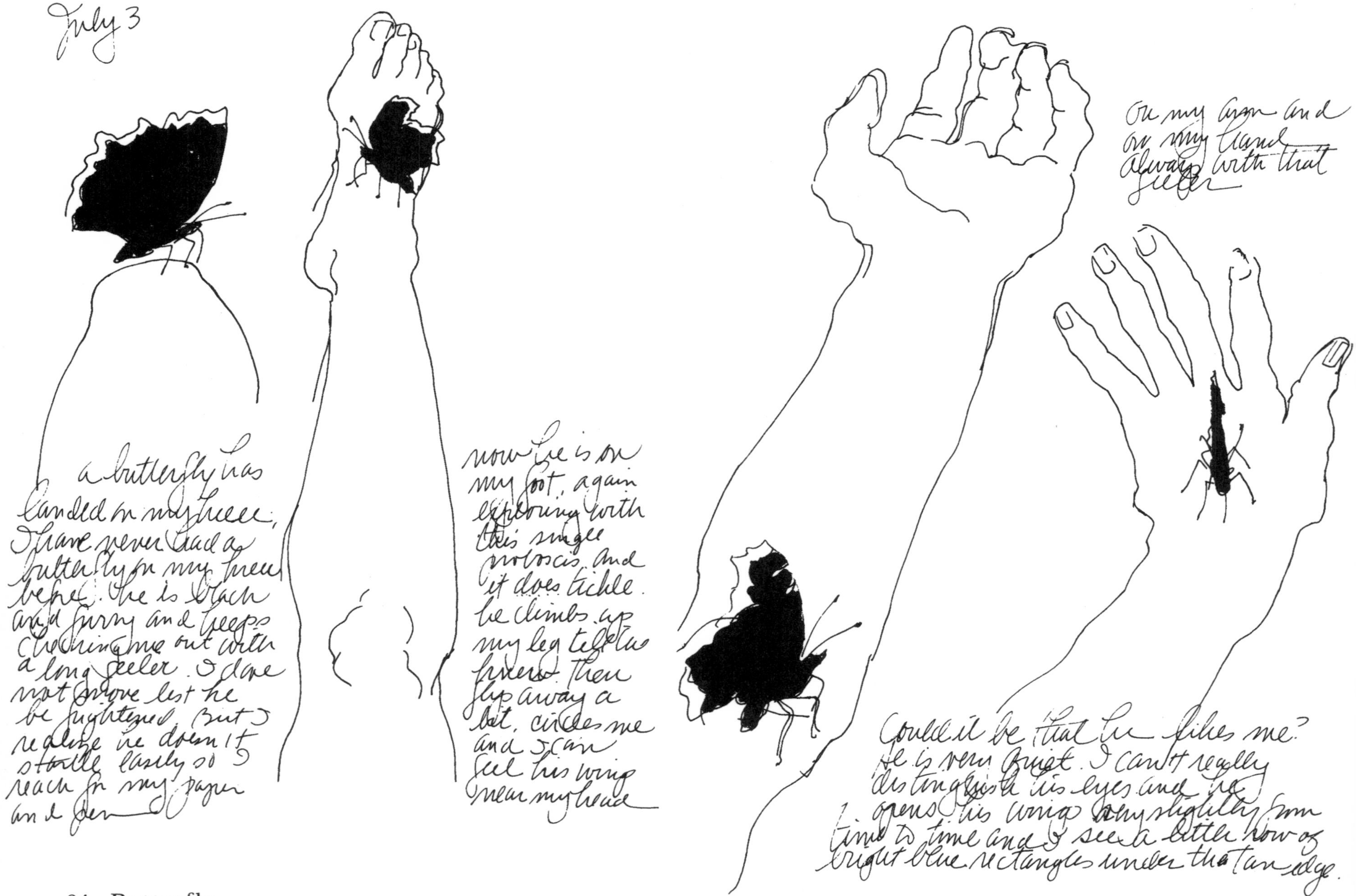

July 3

a butterfly has landed on my knee. I have never had a butterfly on my knee before. He is black and furry and keeps checking me out with a long feeler. I dare not move lest he be frightened. But I realize he doesn't startle easily so I reach for my paper and pen

now he is on my foot, again exploring with this single proboscis. And it does tickle. He climbs up my leg till he hovers, then flys away a bit, circles me and I can feel his wing near my head

On my arm and on my hand, always with that feeler

Could it be that he likes me? He is very quiet. I can't really distinguish his eyes and he opens his wings very slightly from time to time and I see a little row of bright blue rectangles under the tan edge.

94 Butterfly

My butterfly is on my chest. But I see it all now. Ah! It isn't that he likes me at all. He seems to be after the tiny drops of moisture on my skin. So all this time he's been interested in perspiration, not poetry...

I got up to let the cat (sunshine) out and the butterfly followed me to the door, making large circles around me. Now when the cat settled down under my chair the butterfly didn't come near, but as soon as old kitty left for a cooler spot, he was back and as attentive as before. Going in a bit later, he flew around the chair and settled down here after checking the top, the pillow, several spots on the canvas, the foot rail. A regular little goldilocks

95

September 16 — this praying mantis has
spent the day hanging around on the
morning glory vine and has moved
18 inches at the most. I have no idea
what he is doing. He doesn't appear to eat
or look for food or sleep or even
come near a flower, but he naps or
meditates or "prays" constantly except
for a brief period of "arm washing".
Now when the top set of arms are
folded together they look quietly green
with a bit of yellow at the last joint,
but during this washing, with what
looks like a few sets of hands around
the mouth (I think they are mandibles)
one sees a striking black and white
"CBS eye" under the arm, and tiny
white spots as well. And it is at this
point that I first see the heart beat
clearly along the greyish inset on the
abdomen and at a rate of about
one beat for every two of mine.

this may be a vole. I
cannot determine eyes
or ears without probing which
I find difficult to do. there are
indentations which imply eyes.
the fur is a deep lovely brown
and the underside is white

October 6

A strange, flat yellow-winged insect hangs from the moon flower vine, holding by the mouth another insect which looks like a greenish wasp-ish, yellow jacket. I see the yellow jacket at first. It is very still and appears to be dead. Maybe the yellow jacket at first. It is very still and then I see the yellow-winged insect.

I pick it up by a wing and then I see the yellow-winged insect. I leave the yellow jacket near "yellow-wing". I come back in a few minutes. "Yellow-wing" has an arm of a proboscis or some thing inside the yellow jacket's mouth, and when I disturb him tightly and slowly rotating him, then he holds him by a leg joint and puts the proboscis into his ear. My magnifying glass so close it doesn't seem to disturb him, but I turn the vine to see better and the yellow jacket falls to the ground. Could "yellow-wing" be angry? He stands there and I swear he stamps the vine with one fat yellow arm. I see the yellow jacket and brings him back. He gets hooked over a bit, but perhaps she is considered used goods now and "yellow-wing" walks away. A small green arm aside walks by him, stepping on his foot and is ignored. "Yellow-wing" doesn't go far and I'd like to see if he comes back to the yellow jacket, but I can't wait. It's getting dark and cold and I want my tea.

I always thought it was the birds who ate the tomatoes

Opposite: Praying mantis and vole

97   Yellow jackets

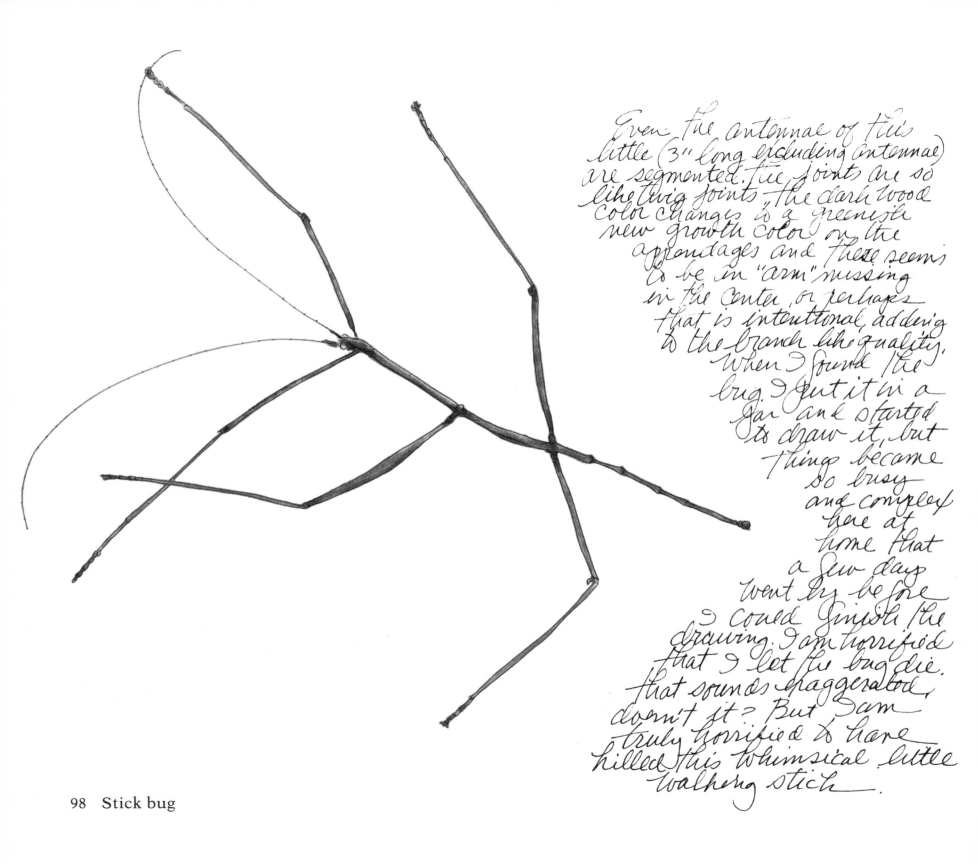

Even the antennae of this little (3" long excluding antennae) are segmented. The joints are so like twig joints, the dark wood color changes to a greenish new growth color on the appendages and there seems to be an "arm" missing in the center, or perhaps that is intentional, adding to the branch like quality. When I found the bug I put it in a jar and started to draw it, but things became so busy and complex here at home that a few days went by before I could finish the drawing. I am horrified that I let the bug die. That sounds exaggerated, doesn't it? But I am truly horrified to have killed this whimsical little walking stick.

98 Stick bug

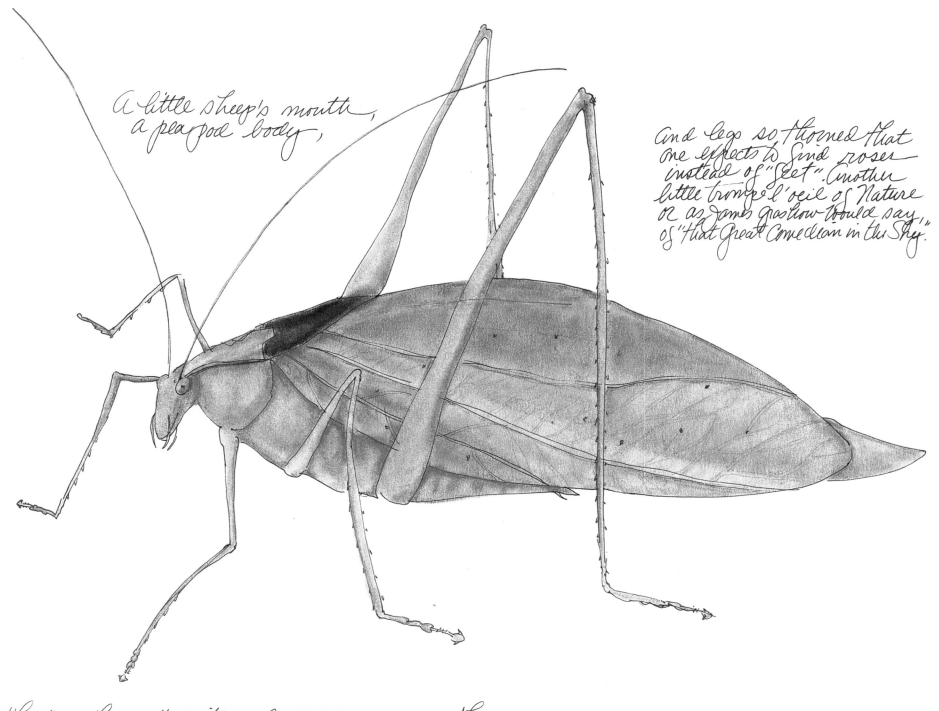

A little sheep's mouth,
a pea pod body,

And legs so thorned that
one expects to find roses
instead of "feet". Another
little trompe l'oeil of Nature
or as James Grashow would say,
of "that Great Comedian in the Sky".

This "Grasshopper" with a leaf for a wing, the plant with a wing for a leaf, the so
many things so like the so many other things, that more and more, I think it is
not that Everything is like something else, but that Everything Is Everything Else.
I see myself in the lettuce and the landscape.

99 Grasshopper

12:00 — June 11. I find 4 worms, perfectly color coordinated to my brussel sprout plant, and wonder just how much one eats so I take one worm and one perfect leaf together in a jar; most appropriately hear the 12:00 whistle; say "lunch time" and leave him to find the leaf. He climbs straight up the stem center, head swaying back & forth

creeps down the entire right hand side of the leaf, then stops and eats, and in 7 minutes has an astonishing "bite" in the leaf, eating from right to left. Stops just as abruptly as he started, and climbs back down the leaf to the center vein, up (drops a small green "dropping"), climbs back up the center very slowly and back to the same eating spot.

1:00 — continues to eat pretty much where he had left off again for about 6 or 7 minutes, right to left, then all the way down the stem to the edge of the jar. looks around a bit (but I'm not really sure if he has eyes) proceeds back up the stem with this sweeping swaying head motion. another dropping

1:30 — up to the top for another bite and another dropping then down to one of the veins. Not moving at all for quite some time until a dropping is positively thrown? But the funny thing is that I'm not sure from which end it was thrown

2:30 — this is the third time back to this spot. I shall spare us all the enumeration of each dropping. I shall draw them and they can be counted, but the pattern seems to certainly be a short fast period of eating, then a stopping and a dropping.)

3:30 — a lot a time spent in trying to eat through the tough center vein. and now a rest along that same vein lower down. He spends a good bit of resting time on that seam and it causes me to wonder if by following that line he is hardly visible.

8:00 — I mean, I know that he is hardly visible on that central vein. I think that he must know it as well, because he uses it as his hide out between eatings. He has 3 sets of feet on the upper 3 segments & 4 on the lower

11:00 — my magnifying glass shows me that he has an amusing little walrus-like face and that it is with a pair of garden shears for a mouth that he eats. The clipping in a left-right opening and closing

8:00 — the next morning, June 12 and he proceeds with the same voracity. You can be certain that I run straightaway to my garden in search of any overlooked brother on my brussel sprouts or cabbages. I meet with two and they meet with death.

Cabbage worm

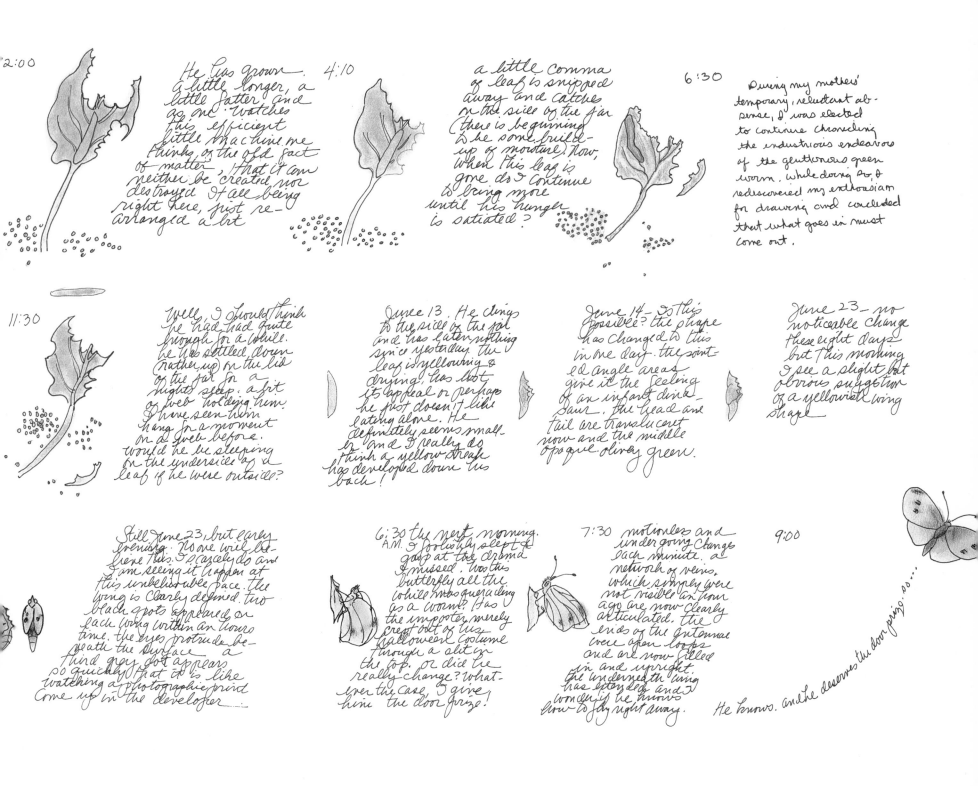

2:00

He has grown. A little longer, a little fatter and as one watches this efficient little machine, one thinks of the old fact of matter, that it can neither be created nor destroyed. It's all being right here, just re-arranged a bit.

4:10

A little comma of leaf is snipped away and catches on the side of the jar. (There is beginning to be some build-up of moisture.) Now, when this leaf is gone do I continue to bring more until his hunger is satiated?

6:30

During my mother's temporary, reluctant absence, I was elected to continue chronicling the industrious endeavors of the gluttonous green worm. While doing so, I rediscovered my enthusiasm for drawing and concluded that what goes in must come out.

11:30

Well, I should think he had had quite enough for a while. He has settled down (rather up) on the lid of the jar for a night's sleep. A bit of web holding him. I have seen them hang for a moment on a web before. Would he be sleeping on the underside of a leaf if he were outside?

June 13. He clings to the side of the jar and has eaten nothing since yesterday. The leaf is yellowing & drying, has lost its appeal or perhaps he just doesn't like eating alone. He definitely seems smaller and I really do think a yellow streak has developed down his back!

June 14 - Is this possible? The shape has changed to this in one day. The pointed angle areas give it the feeling of an infant dinosaur. The head and tail are translucent now and the middle opaque, olivey green.

June 23 - no noticeable change these eight days but this morning I see a slight but obvious suggestion of a yellowish wing shape.

Still June 23, but early evening. No one will believe this. I scarcely do and I am seeing it happen at this unbelievable pace. The wing is clearly defined. Two black spots appeared on each wing within an hour's time. The eyes protrude beneath the surface. A third grey dot appears so quickly that it is like watching a photographic print come up in the developer.

6:30 the next morning. A.M. I foolishly slept & gasp at the drama I missed. Was this butterfly all the while masquerading as a worm? Has the impostor merely crept out of his Halloween costume through a slit in the top. or did he really change? Whatever the case, I give him the door prize!

7:30 motionless and undergoing changes each minute. A network of veins which simply were not visible an hour ago are now clearly articulated. The ends of the antennae were open loops and are now filled in and upright. The underneath wing has extended and I wonder if he knows how to fly right away.

9:00

He knows. And he deserves the door prize...

Sometimes a large bee will follow our loud buzzing lawn mower. He will come so close to alighting upon this great green thing. Does he think it is some ally or enemy? Or queen bee? I can understand this interest. But a tree frog? Once, a tree frog persistently followed behind the lawn mower for a distance of at least 50 feet. He shirted back and forth behind the

KEEP HANDS AND FEET FROM UNDER MOWER

flap. Do other tree frog sound like our Lawn Boy? Or is there something about trees that I don't know?

THE WILD GARDEN

The wild garden is not something you can stand and face. It hits you hale in the back and the front at the same time. I step beyond my tame gardens into the woods and I step into a surprise party. Behind every trunk something lies waiting to spring surprise. Is nature all calls and answers and no pauses? It seems there is no place where something has not invited itself to grow. I ask how the lichen on the surface of a rock and the violet in the crevice can possibly earn a living there. I keep forgetting that they want to grow. It's their lives I'm asking about.

I walk through the woods, opening "presents" all the way to the river. I watch the game of catch the river has with the sky: it catches the light and throws it back, catches it again and throws it back. I stop in a field to pick blond grasses in huge bouquets. Am I trying to build a field in my living room? What I want is a field outside my window, a flag of a field I can wave back to.

I once tried to create a field. It was folly of me to think that I could. A sneaking carnival instinct made me plant an unnatural profusion of wildflowers. I need to learn from nature's "light" of hand.

105 Wild grapes

June 14 Wild grapes - Old Church Lane
I am marking this vine with
a ribbon of yarn so I can come back again
and find it in among all the others.
It is 17½" from the ribbon to the right hand
tip and there are three bunches of
light green buds.

July 1
Not really much
length growth, but the leaves are
larger and the grapes are really
developing. If you want to make
stuffed grape leaves, this is the time to
pick them before they get too tough.

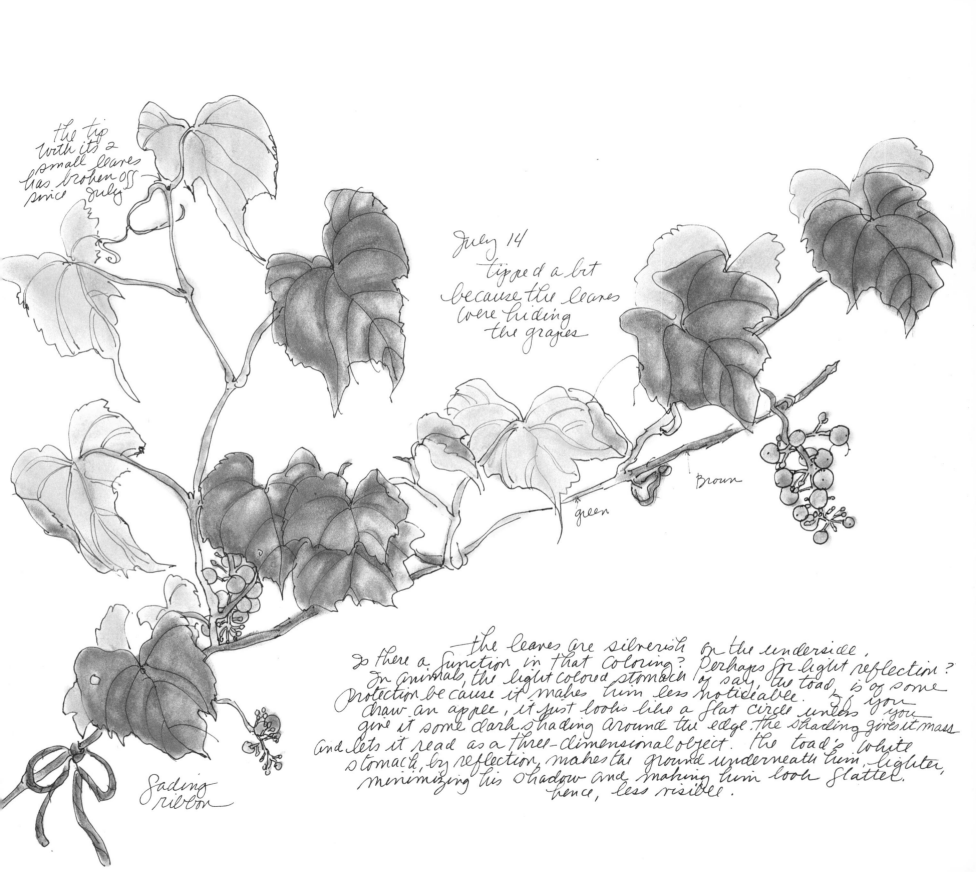

the tip
with its 2
small leaves
has broken off
since July

July 14
tipped a bit
because the leaves
were hiding
the grapes

green

Brown

fading
ribbon

the leaves are silverish on the underside.
Is there a function in that coloring? Perhaps for light reflection?
In animals, the light colored stomach of say, the toad, is of some
protection because it makes him less noticeable. If you
draw an apple, it just looks like a flat circle unless you
give it some dark shading around the edge. The shading gives it mass
and lets it read as a three-dimensional object. The toad's white
stomach, by reflection, makes the ground underneath him, lighter,
minimizing his shadow and making him look flatter,
hence, less visible.

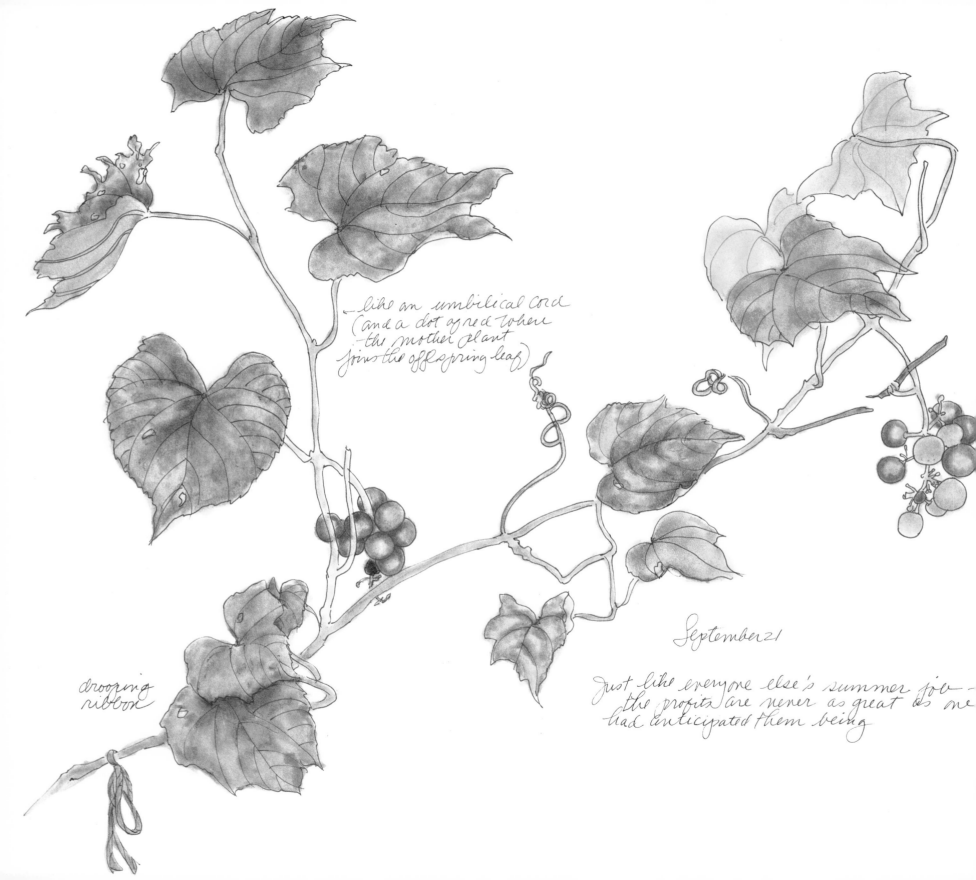

— like an umbilical cord
(and a dot agreed where
the mother plant
joins the offspring leaf)

drooping
ribbon

September 21

Just like everyone else's summer job —
the profits are never as great as one
had anticipated them being

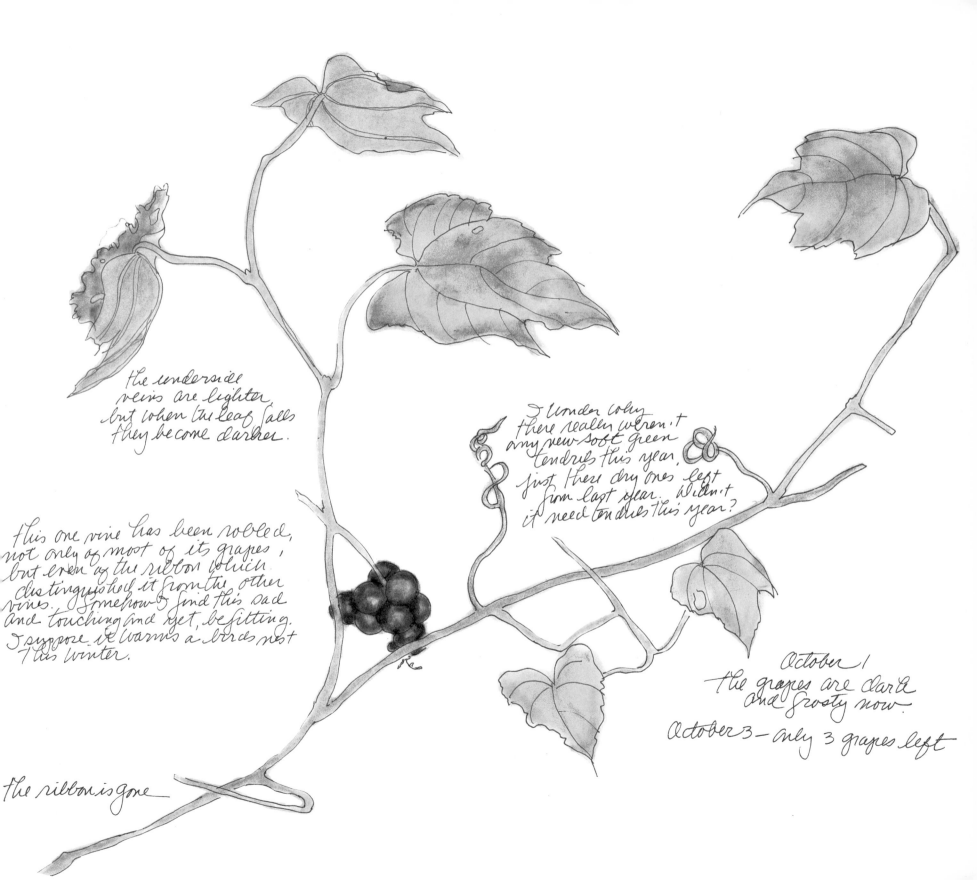

the underside
veins are lighter,
but when the leaf falls
they become darker.

I wonder why
there really weren't
any new soft green
tendrils this year,
just these dry ones left
from last year. Wasn't
it need tendrils this year?

This one vine has been robbed,
not only of most of its grapes,
but even of the ribbon which
distinguished it from the other
vines. Somehow I find this sad
and touching and yet, befitting.
I suppose it warms a birds nest
this winter.

October 1
the grapes are dark
and frosty now.

October 3 — only 3 grapes left

the ribbon is gone

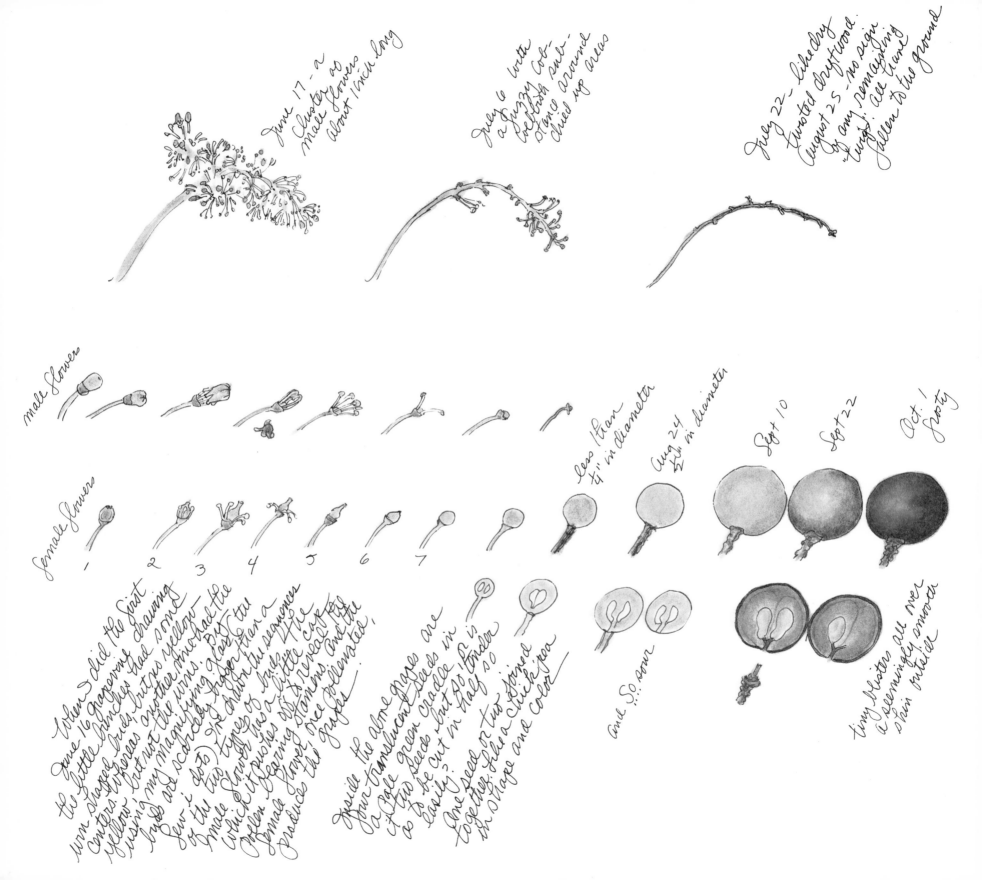

June 17 - a cluster of male flowers about 1 inch long

July 6 with a fuzzy cob-webbish substance around - dried up areas

July 22 - like dry twisted driftwood. August 25 - no sign of any remaining twigs; all have fallen to the ground

male flowers

female flowers

1 2 3 4 5 6 7

less than 1/4" in diameter

Aug 24 1/2" in diameter

Sept 10

Sept 22

Oct. 1 spotty

are so sour

I

III

leoves

This elegantly curved tip could so easily go unnoticed as it is just 2½ inches from I to II and one inch from II to III. The folded leaves are as fuzzy and as frail in color as a baby blanket and the open leaf has a hook tatted edge

So that's how it works! You take your drawing pad and go down to the end of Arthur's driveway where the wild grapes grow, almost as far as Old Church Lane. You sit on the overturned soup pot in the midst of the vines and tall grasses. You are drawing the vine when you hear some passersby discussing whether or not they should walk up the hill to look at Arthur's garden. You try to squeeze deeper into the vines so that they won't see you as they pass, but this is a little strange, isn't it? What if they see you anyway? That would be stranger. You'd better make your presence known because even if they don't see you on the way up, they probably will on their way down.

You stand up and gaily say hello and that you hope you haven't frightened them, and that what you are doing here is drawing the grapevine. At this point there are just a few dry leaves and grapes left, so you realize that they are wondering why ever you are drawing them. They proceed, out of embarrassment perhaps, to tell you a little local gossip (and you know that you've just become an addition to that gossip). In parting, they tell you how many neighbors are selling their houses and for just how much.

That's how it all begins. You take your drawing pad down to the end of the driveway and wait to see what you hear by the grapevine.

113

December 20   I have cut the vine and bound it round.
For the ribbon lost, another found.

Blackberries are so old-fashioned, so unsophisticated. They are the remembrance-of-things-past berry. They grow along dusty lanes and are their own barbed wire fences.

When I was a child there was a blackberry patch in a lot not far from our house. It was our custom each year to spend one summer day picking the berries for jam. Now this little foray had to be planned well in advance as those were the war years and many things were rationed. Months ahead of the preserving season my mother would begin trading in meat and shoe coupons for sugar. When the berries were ripe she and Mrs. Davis would put on their picking costumes—my mother's was a red-and-white polka-dotted blouse and a pair of light blue ladylike overalls, the kind you see in old photographs of women working in defense plants—and we'd head off with our pails.

I liked hanging the little wire handle of my berry pail over my wrist so that I could hold a branch with one hand and pick with the other. And it was nice to be with my mother and Mrs. Davis. But the sun is harsh on a fair child and it would send me home early. They'd come home rosy and scratched and late and with enough berries for jam for everybody's Christmas.

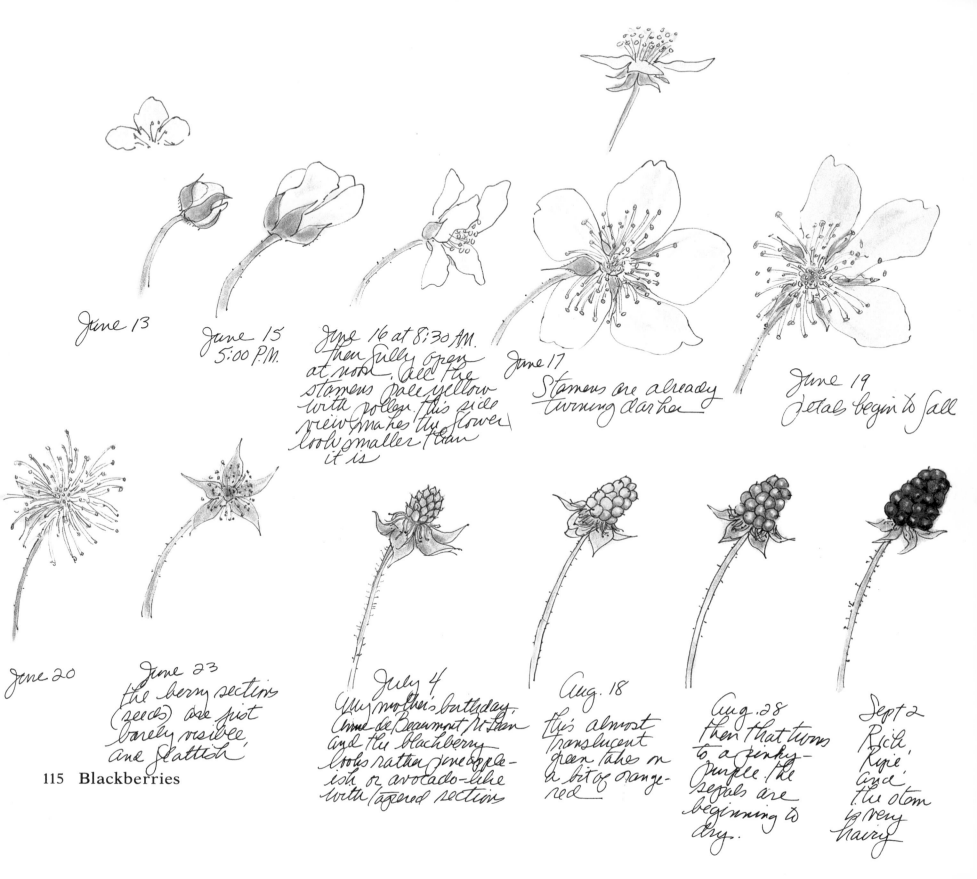

June 13

June 15
5:00 P.M.

June 16 at 8:30 A.M.
then fully open
at noon. All the
stamens pale yellow
with pollen. This side
view makes the flower
look smaller than
it is

June 17
Stamens are already
turning darker

June 19
Petals begin to fall

June 20

June 23
the berry sections
(seeds) are just
barely visible,
and flattish

115 Blackberries

July 4
my mother's birthday.
Anne de Beaumont McLean
And the blackberry
looks rather pineapple-
ish or avocado-like
with tapered section

Aug. 18
this almost
translucent
green takes on
a bit of orange-
red

Aug. 28
then that turns
to a pinky-
purple. The
sepals are
beginning to
dry.

Sept 2
Rich,
Ripe,
and
the stem
is very
hairy

June 7

August 15

5:51 A.M.

118   Bindweed (wild morning glories)

12:21 P.M.

Wild
Morning Glorie
(Morning Glory)
(Bindweed)
on Bear Island
Maine

5:22 P.M.

7:03 P.M.
The tips of the blossoms
have become ever so
slightly rusted

3:18 P.M.
the next day

seed
pods

I am not as yet at one with nature,
nor even at one with lettuce,
but I am a little closer.
Things are beginning to join hands now.
I am going to keep paying attention
to this good green earth.
I am just learning
to tap my foot in time to it.